S0-AFE-294

"*Mind Whispering* is the connection to a paradigm shift in consciousness. Tara Bennett-Goleman's poetic storytelling is addictive, mesmerizingly beautiful, and meditatively calm. *Mind Whispering* is a little gem, a story of conscious evolution."

—Dr. A.T. Ariyaratne, founder and president of the
Sarvodaya Movement

"An extraordinarily beautiful, compassionate exploration of the nature of our minds, the roots of our suffering, and a clear pathway toward healing. If you struggle with habits of mind that prevent you from being fully alive, this book can free you."

—Mark Hyman, M.D., author of the *New York Times* bestsellers
The Blood Sugar Solution and *The UltraMind Solution*

"With a masterful amalgam of clinical wisdom, modern neuroscience, and ancient Buddhist psychology, Tara Bennett-Goleman's *Mind Whispering* will teach you how to harness the light touch of a horse whisperer to shed negative emotional habits, raise your positivity ratio, and thereby flourish."

—Barbara L. Fredrickson, Ph.D., author of *Positivity* and
Love 2.0

"In weaving together so many different strands of wisdom and compassion, *Mind Whispering* creates an inspiring tapestry of engaged understanding. Tara Bennett-Goleman draws on an amazingly rich and varied life and in so doing illuminates our path of love and freedom."

—Joseph Goldstein, author of *A Heart Full of Peace* and
One Dharma

"Tara Bennett-Goleman's *Mind Whispering* furthers the dialogue of psychology east and west. She is truly a pioneer in the integration of cognitive therapy and mindfulness. This book is not only useful for clinicians, but for people in their everyday lives."

—Dr. Aaron Beck, founder of the Beck Institute for Cognitive
Behavior Therapy

"My dear friend Tara shares her wise and playful insights into why we are continually at the mercy of negative mind patterns and how, if we do the work, we can change them. Her longtime Buddhist and therapeutic trainings bring fresh insights through delightful wisdom stories and clear common good sense. *Mind Whispering* is like spending time with a sane and cherished friend, whose warmth and wisdom we can completely trust."

—Richard Gere, actor and activist

"Breathtaking in scope, profound in its message, and inspiring in its invitation, *Mind Whispering* has something for everyone. Tara Bennett-Goleman is a psychotherapist, meditator, student of horse whispering, science synthesizer, and arts practitioner who brings all of these extraordinary modes of knowing to understanding the subtle habits of our mind and how we can flourish. This book is brilliant and will be helpful to all!"

—Richard J. Davidson, Ph.D., neuroscientist at the University
of Wisconsin and coauthor of *The Emotional Life of Your Brain*

"*Mind Whispering* is an inspired and innovative look at the ways we can be tripped up by our habitual thinking. As she did in *Emotional Alchemy*, Tara puts a refreshingly clear and practical take on the subject, this time beautifully melding the teaching of Buddhism, horse whispering, and her own therapeutic practice."

—Bonnie Raitt, musician and activist

"Tara Bennett-Goleman has done a masterful job in *Mind Whispering*, synthesizing ancient wisdom practices, modern cognitive psychology, and 'horse whispering.' With passionate concern and personal experience, our guide focuses on how 'modes' of being influence our perceptions, thoughts, and emotions. She demonstrates how modes shape our sense of reality in often hidden ways that repeatedly constrict our emotional lives. These new insights offer empowering suggestions on how to liberate ourselves from automatic modes, such as insecure states of mind, as we learn to create more freedom, kindness, and compassion in our inner and interpersonal lives."

—Daniel J. Siegel, M.D., author of *Mindsight*, executive director
of the Mindsight Institute

"Tara Bennett-Goleman has written a book of exquisite sensitivity and wisdom. With a new map to guide, she invites us to cultivate a healing relationship with the deepest aspects of ourselves and the world by brilliantly synthesizing the ancient wisdom of Asia with the findings of modern psychological science. It will be used not only by meditation teachers, clinicians, and therapists, but also by all who are seeking a new way to find courage and hope in their lives."

—Mark Williams, professor of Clinical Psychology at the
University of Oxford and coauthor of *Mindfulness*

MIND WHISPERING

A New Map to Freedom from
Self-Defeating Emotional Habits

Tara Bennett-Goleman

HarperOne
An Imprint of HarperCollins*Publishers*

HarperOne

MIND WHISPERING: *A New Map to Freedom from Self-Defeating Emotional Habits.* Copyright © 2013 by Tara Bennett-Goleman. All rights reserved. Printed in the United States of America. No part of this book may be used or reproduced in any manner whatsoever without written permission except in the case of brief quotations embodied in critical articles and reviews. For information address HarperCollins Publishers, 10 East 53rd Street, New York, NY 10022.

The excerpt on page 184 is from the story "Northern Lights Dancing—The Naming of a Tea Bowl," copyright © 2011 by David Street. Used by permission of the potter.

HarperCollins books may be purchased for educational, business, or sales promotional use. For information please e-mail the Special Markets Department at SPsales@harpercollins.com.

HarperCollins website: http://www.harpercollins.com

HarperCollins®, ♣®, and HarperOne™ are trademarks of HarperCollins Publishers

FIRST HARPERCOLLINS PAPERBACK EDITION PUBLISHED IN 2014

Designed by Level C

Library of Congress Cataloging-in-Publication Data

Bennett-Goleman, Tara.
Mind whispering : a new map to freedom from self-defeating emotional habits /
Tara Bennett-Goleman. — 1st ed.
p. cm.
Includes bibliographical references and index.

ISBN 978–0–06–213131–7

1. Meditation. 2. Emotions. 3. Calmness. 4. Peace of mind. I. Title.
BF637.M4B463 2013
158.1'2—dc23
2012029111

14 15 16 17 18 RRD(H) 10 9 8 7 6 5 4 3 2 1

To the light of wisdom and the heart of compassion in everyone

CONTENTS

Foreword   ix

PART ONE
Modes of Being

One   The Lotus Effect   3

Two   The World of Modes and Why They Matter   13

Three   Root Causes   25

Four   Insecure Connections   37

Five   An Evolutionary Arms Race   47

Six   Traps, Triggers, and Core Beliefs   59

Seven   The Evolution of Emotion   69

PART TWO
Mind Whispering

Eight   Shifting the Lens   87

Nine   The Art of Whispering   99

Ten   The Mindful Overseer   113

Eleven   Mode Work   129

Twelve   Priming Our Secure Base   149

Thirteen   Training the Mind   165

Fourteen   Wise Heart   179

Fifteen   The Physics of Emotion   195

PART THREE
Tending to the Whispers of the World

Sixteen   Two of Me, Two of You   221

Seventeen   Joined at the Heart   235

Eighteen   A Shared Secure Base   245

Nineteen   The Transformative Power of the Arts   261

Twenty   Collective Whispering   273

Twenty-One   Connected at the Source   289

              Acknowledgments   299

              Resources   303

              Notes   305

              Index   317

THE DALAI LAMA

Foreword

We all want to be happy, but in today's world we may be confused about how to achieve it. In my own experience and that of the author of this book, the source of real and lasting happiness is in the mind. The key to finding happiness and overcoming problems is inner peace, the source of which is not to be found in sensory pleasures, delightful though they may be, nor in physical exercises, but in working with the mind.

We may not see it at first, but many of our problems are our own creation, and what I find encouraging about this is that it means that solutions to them are also within our reach. Pacifying the mind is not easy; it takes time and hard work, but that is true of any human endeavour. You need determination right from the start, accepting that there will be obstacles, and resolving that despite them all you will continue until you reach your goal.

This is not to say that in transforming the mind pressure or force are involved. It is something that needs to be approached voluntarily and willingly. Tara Bennett-Goleman has drawn from several different sources in preparing this book, among them Buddhist teachings she has become familiar with and the therapeutic insights of our mutual friend Dr Aaron Beck. But what she has learned from working with a horse whisperer, and which gives the book its title, is the importance of becoming attuned to the mind's needs and concerns, approaching problems and finding solutions to them with sensitivity and intelligence.

I have no doubt that readers, whether they wish to address their own every day problems or seek to help others deal with theirs, will find a great deal in this book to delight and inspire them.

His Holiness the Dalai Lama
December 31, 2012

PART

Modes of Being

CHAPTER

The Lotus Effect

A vast river flows through Bangkok, dividing the city's sprawl. From my hotel window I watch elegant, handcrafted teak boats glide through the undulating waves. At the shoreline I notice what appears to be a cluster of lotus pads bobbing up and down with the rhythmic swells.

On closer examination, I see lotus pads surrounded by an eddy of garbage, the random detritus in the river. Their lotus flowers blossom anyway, rising above the debris.

Fascinated, I begin to reflect on how we human beings have the same capacity as those lotus flowers, which in Eastern cultures are symbols of enlightenment. They teach that we can rise up out of our own form of debris and muddiness, those painful ways of being that can burden our relationships, our work, and our inner lives.

Mind whispering means attuning to the subtle habits of our minds and hearts, to uncover the qualities deep within us that can allow wisdom to bloom lotus-like out of the mud of confusion.

Learning to see the ways in which we succumb to the muddiness of our minds over and over again is a first, crucial step. My friend Steven recently told me about the anguish that had plagued him for close to a year as he watched his beloved son prepare to go off to college. A single father, he was wholeheartedly dedicated to raising his child.

Steven anticipated profound grief at his son's absence, but he was determined to feel happy for his son, who was elated to be spreading his wings. So a few months in advance of his son's departure Steven made it a daily practice to sit quietly and connect with the joy that his son felt. He used a method he'd learned from a Buddhist teacher, called "sympathetic joy"—feeling joy in the happiness of others.

At first, Steven said, it had seemed a bit artificial because this wasn't how he felt most of the time. But as he had reflected and practiced each day, his self-focus started to wane and he was more able to empathize with his son and share in his joy. By the time his son started college, my friend was genuinely happy for him rather than mired in attachment and fear of loss.

The painter Magritte, commenting on his mysterious surrealist images, saw his creative works as "the best proof of my break with the absurd mental habits that take the place of an authentic feeling of existence." Mind whispering seeks to bring our "absurd mental habits" into awareness, to help us attune not just to ourselves but also to others, which creates a warm chemistry—as my friend did with his son.

This work helps us see ourselves as we truly are instead of through the lenses of our emotional habits and mental patterns. Mind whispering is both an educational and therapeutic model, an approach to our minds that seeks to transform our emotions and connect us to our inner wisdom.

There's a Tibetan term, *sanje*, which translates as "waking up and blossoming," akin to how a lotus grows from the mud. As these qualities of the mind and heart awaken, they allow wisdom and compassion to fully bloom. My friend had awakened and blossomed, freeing himself from a troubling mode of being.

What allows the lotus to grow out of the mud? What are the inner qualities that allow wisdom to bloom lotus-like out of a mind of confusion and connect us more genuinely with ourselves and with each other? And how can we break free from modes of being that capture and constrict us?

The "lotus effect" in biology refers to that plant's remarkable waterproof capacity to grow through the mud while remaining pristine. Nothing sticks to lotus leaves. The secret of the lotus's self-cleaning property is a leaf surface filled with tiny bumps, which meet droplets at an angle so only a tiny portion of a water drop contacts the leaf. The drops stick more strongly to particles on the leaf than to the leaf itself, thus cleaning impurities as they roll off.

Like the lotus effect's non-stick properties, the basic practices of mind whispering create a field of non-clinging in the mind. As we'll see, this application of kindheartedness, calmness, and clear awareness allow for a lightness of being.

Patterns That Connect

Bob Sadowski (aka "RJ" Sadowski) is a gifted natural horseman, and I've been studying with him for several years. He practices what he calls horse*mind*ship. Under a federal land-management program, Bob domesticates wild mustangs to prepare them for adoptive homes, yet he uses no force in taming them. In essence, Bob befriends the horse, becoming a trusted member of its herd—even its leader. He knows how to approach a prey animal: with respect. He invites them while also giving them space.

Horses are always inviting us into the present, always ready to connect with us the moment we attune to them. They live in the moment, waiting for us to find our way there too. One day Bob, my horse Sandhi, and I were in what we call the playpen, a round, fenced training ring at the stable where Sandhi boards. Bob was giving me guidance on an energy located in our abdomens, which when pointed toward a horse acts as a force to which the horse is keenly attuned.

He demonstrated by asking Sandhi to move forward, but not with words. He directed what he calls our "core energy" toward the horse. When he turned away, she stood still. But as soon as he turned toward her, she moved forward.

He then guided me to relax and retreat—that is, move my core away from her—and let my energy settle downward. This tells her that she has responded to the communication and engaged in the conversation, which seems to delight her. For horses, engaging and retreating like this is a way of connecting in a language they understand.

Then we asked her to move around the ring with an unbroken continuity, simply by guiding her when needed, letting her learn on her own to respond to these minimal requests, and retreating again to acknowledge her having understood. She ran around the ring with a steady fluidity. Then, as Bob and I stood in the center of the circle, we turned, silently calling her toward us. She broke her step and trotted over to us. And then she lightly rested her head on my shoulder, linking us in the equivalent of a spontaneous warm hug. I felt an amazing connectedness. When I started to walk around the ring, she stayed glued to me, her head gently touching my back. I felt like one of those mythic half-human, half-horse creatures—a centaur.

In horse whispering, this deep connection is called "joining up," and we cherish it in our lives whenever and however it occurs. In those moments, any sense of separateness dissolves and it's like we are one being, replete and perfectly content in a shared cocoon. There's an invisible link.

The ways we humans act and think of ourselves as separate and in control of things must, to a horse, appear strange, even predator-like. But horses seem to accommodate our foolish ways, accept us anyway, and even find creative ways to remind us that we're really part of the herd—as though we've just temporarily forgotten. They are always ready to join up with us; it's where they live.

Joining up expresses our interconnection; it gives us a direct experience of natural relatedness. As with a horse, whether we notice or not, it's always there, waiting for us.

Such connections can arise when we feel a deep and genuine bond with another being. But they can also occur spontaneously in any number of ways—for instance, through creative absorption, from the inspiring beauty of the natural world, or in meditative immersion. When we shift into such a flow of being, we embody a pattern that connects. A deep sense of wellbeing, security, and receptivity pervades our hearts. In this way of being, we yield to our positive qualities, perform at our peak, and open ourselves to a deep resonance with others.

When I have glimpses of this connective force, and a sense of what's possible, I wonder: Why are we settling for less? My heartfelt intention in writing this book is to help free us from patterns that obscure how interconnected everything already is naturally, to melt the barriers to the patterns that connect us.

Patterns of disconnection interrupt this flow of being. The horse-whispering tradition sees such disconnection as relics of the ancient predator–prey dance played out for ages in the equine world. It exists more subtly in the human realm. In these disconnecting patterns, we may find ourselves feeling insecure, holding to distorted views, acting in self-focused or dysfunctional ways, and tuning out. Such states are like a fog of bewilderment moving through the mind. Like fog, these modes of being are not fixed, solid parts of us, but passing conditions. That perspective is at the heart of this work.

When we fail to recognize the bewilderment created by these disconnecting emotional habits and their filters on our lives, the fog can settle over us. But just as the sun evaporates clouds to reveal a clear sky, these inner fogs can dissolve when we recognize them for what they are, see their transparency, and reconnect with our natural clarity.

A New Lens on the Mind

Broadway director David Cromer captures the way our bewildering modes distort our sense of the world: "I was easily defeated, easily shut down, easily insulted. If the train door closes right before you get into it, you go, *Oh, I missed the train.* What I would do was, *The train hates me. I don't deserve to get on the train.*"[1]

Cromer was describing the world seen through the window of depression, which Dr. Aaron Beck, the founder of cognitive therapy, sees as an extreme mode. He proposes that the mode concept offers us a way to rethink what ails the mind and how to help.

I was partly inspired to delve into modes by Dr. Beck's writings.[2] My earlier work focused on deep emotional patterns. But modes occur in a much wider range of our experiences; we are always in one mode or another.[3]

Modes are distinct orchestrations of how we feel and think, what we desire and where we focus our attention, what we perceive and how we behave.[4] Some modes open us to delight and wisdom; others close us down into fear or, like David Cromer, despair and self-pity. Modes are like invisible puppeteers of the mind, pulling our strings while hiding backstage. We rarely realize how a mode drives and distorts our experiences.

Our mode dictates what we notice and what we do not, and so creates our subjective world.[5] Every mode gives us its unique lens on our lives; the more distorted that lens, the more negative the mode. Can people always be trusted, or never? Are we confident we're up to life's challenges, or fear we'll be overwhelmed? Can we roll with the punches, or are we easily insulted, shut down, defeated—feeling we do not deserve to get on that train?

Broadly speaking, there are three varieties of modes: the maladaptive, where our dysfunctions rule; the adaptive zone, where we are at our everyday best; and a further spectrum that moves us toward a lightness of being.

We'll look at several modes in this book, sampling the spectrum of models from East to West. Although this by no means maps the whole of human experience, combining these maps offers a more complete picture of our possibilities, from bewildered to wise.

The second part of this book details mind whispering, a set of perspectives and practices for transforming modes through what I call "mindful habit change." The major strategies for handling negative modes are to avoid them by shifting to a more positive range, transmute them through mode work, or transcend their negative qualities and strengthen their positive elements. Mind whispering adapts this triad of approaches.

Finally, at the collective level we'll explore how partners, families, groups, communities, and nations can share a way of being that guides perception, understanding, and action. In perhaps the most destructive collective mode, a shared insecurity feeds a too-narrow sense of "us" and gives rise to a collective hostility toward "them," a group perceived as an enemy. Healing such divides can begin by inviting all parties into a shared secure mode. Emotions are a force that can separate or connect.

Unseen Forces

In the San Juan Islands of the Pacific Northwest, I was waiting for a ferry to an island oasis. I had settled on the pier and was absorbing the gentle beauty of the natural world: lavender and cobalt tones illuminated the clear water and pristine atmosphere. A gentle opening inspired a refined, clear awareness.

As my gaze rested softly on the peaceful waters, I noticed an area near the ferry dock where a blanket of oil was spreading across the surface of the sea's natural purity. My heart sank as my thoughts swirled into a very different mental landscape, the poignant reminder of the damage being done to the natural world by forces of greed and ignorance. The image of that disturbing oil slick seems an apt metaphor for how unseen forces fuel the root causes of our destructive modes, whether they overpower the purity of the environment or the clear awareness of our own minds. We need to address these root causes in our efforts to restore balance in our relationship with our natural home, and with our minds.

For me, the layer of oil graphically symbolizes the natural cause-and-effect sequence at the heart of Buddhist psychology, which describes how our pure, open, and clear awareness becomes obscured by bewilderment. Unaware of what drives us and where we are headed, we act in ways that disconnect us from the web of all things. The optical illusion of a separate self asserts itself, impervious to the consequences of not seeing how things naturally are, imposing arbitrary preferences on a harmonious natural order. When billions of the world's beings impose their various likes and dislikes, it adds up to the world as we know it today.

Picture how the Parthenon looms large over the Athens skyline, lit up on a high hilltop, a backdrop to the urban jumble of neon signs, billboards, and skyscrapers. Yet amidst this mix of ancient and modern, the Parthenon holds its own. It's the womb of Western civilization and has maintained a presence over the centuries in this now-bustling modern city.

As a visual metaphor for the ancient within the contemporary, the Parthenon's continuing prominence—a reminder of the philosophical dialogues that once rang near its halls—resonates with some perennial truths about human behavior so primal that they must be set in the very wiring of our brains. We're drawn to that which we find pleasing and are repelled by that which displeases. We are open to that which inspires and closed to that which frustrates. We reach out to those we are comforted by and turn away from those we are challenged by.

These are clearly natural inclinations. But how then do we evolve and grow beyond self-imposed limitations, so that we learn from that which we're challenged by, and so our desires fuel our passion to care? These are among the queries that guide our journey here.

Evolving Emotions

As a seasoned psychotherapist and teacher on the one hand and a long-term meditator on the other, I have brought the best of my learning into developing the mind-whispering approach. Mind whispering melds both ancient and contemporary methods, taken from both the East and the West and gathered from a wide range of learning environments.

Over the years I've attended meetings organized by the Mind and Life Institute, where the Dalai Lama delves into science topics with groups of researchers. I've received teachings and had interviews with meditation masters from Tibet, Nepal, India, and Burma. Put that together with my training as a cognitive therapist, the many hours I've spent in pastures studying with a horse whisperer, and my concern for and involvement in environmental and social issues.

Within these many varied worlds I began to recognize patterns, with similar themes in different settings. I realized how one system had much to offer others. When I was in one situation, I'd think, *Some principles from horse whispering could really be useful now*, or *Scientists might be interested in exploring the Buddhist practice of investigation*. I networked ideas and approaches, synthesizing and trying to translate practices and principles into an accessible language, weaving these many threads into the fabric of mind whispering.

I began my studies of meditation principles in India, where I became a student of the ancient system of psychology at the heart of Buddhism, in which the "therapy" is meditation. This psychology has a refreshingly positive view of human nature. It recognizes our disturbing mind states but sees them as temporarily covering our true nature, like clouds covering the sun.

When I returned to the West and studied counseling psychology, this Eastern perspective was my reference point, and it allowed me to learn about dysfunctional patterns without seeing people as being defined by them. Throughout my work in psychology I've stayed focused on this West–East fusion.

Western psychology has concentrated on how to heal maladaptive patterns and, more recently, how to foster positive ones. Eastern views offer a wider horizon, mapping an additional landscape—a flourishing range that Western psychology is just beginning to recognize, and one that transcends the adaptive range.

The Eastern approach uses methods for deep habit change, all of which aim to free us from conditioned patterns by awakening a clear mind and kind heart. As I practice each day toward transforming my own mind, and continue to study, I see more clearly how each of these outlooks says some of the same things, but in a different language, and how each can complement and refine one another.

Some Asian meditation teachers have told me that they are perplexed by the emotional turmoil of their Western students and have found that some contemporary psychological systems help with this. Meanwhile, Eastern perspectives have much to offer psychology, adding methods for freeing the mind from the varieties of mental bondage.

In addition, horse whispering brings another set of inner tools, including the art of working together across species, a divide that demands fine attunement to an alternate universe of experience. These arts can be adapted to help us navigate our inner turmoil, attune to the mind's subtle currents, and apply gentle guidance—natural principles that fall into place through the connective force of joining up.

Another source of mind whispering draws on new findings in neuroscience. This research shows how our most negative modes—which can fill us with overwhelming emotions, such as dread or rage—are driven by the limbic part of our brain. As we shift from this negative mode range into the positive, there is greater control by the executive center as our disturbing emotions give way to positivity and our focus frees up from fixations.

Mind whispering blends perspectives and practices derived from a large and inclusive Eastern-and-Western toolbox. The result: a mix of ancient psychologies and their methods with contemporary therapeutic tools grounded in current scientific findings.

My sincere hope is that this approach offers a path to evolving our emotions, individually and collectively. Mind whispering makes our minds more clear and free, our perceptions more true, our responses more artful, our connections more genuine, and our hearts more joyous.

The road in life forks at every moment, with one path leading toward confusion, separateness, and entanglement, and the other toward clarity, connection, and mental freedom. With mind whispering, the choice can be ours.

There's a Native American story in which a grandfather tells his grandson a story about two wolves fighting in his own heart. One wolf is vengeful and violent; the other is loving and compassionate.

The grandson asks, "Which wolf will win?"

The grandfather says, "The one I feed."

The World of Modes and Why They Matter

My stepson Hanuman went on a vacation with a girlfriend some years ago. A longtime musician and songwriter, he was accustomed to bringing a guitar along on trips, to take advantage of the open time when inspirations might come.

His smallish travel guitar, a bit beaten up, has never been a problem; it's fitted into overhead luggage compartments wherever Hanuman has gone in the world. He once took it to India and back. But this one time a guy at airport security refused to let him take it through, saying, "The rules don't allow it. You'll have to check your guitar as baggage"—a recipe for disaster since the guitar had no case.

Hanuman tried to explain that he'd always been able to take it aboard and store it in the overhead with no problem. But the guy wouldn't budge.

This triggered a rebellious streak in Hanuman and the two locked horns. Neither of them would give in, and their encounter spiraled downward. Hanuman was angered by how the security guard was fixed in his attitudes and couldn't be open to considering other possibilities. But Hanuman, too, was feeling so reactive that he couldn't, either.

Just at the low point, his girlfriend stepped in, and with utter calm and lots of charm politely said to the security guard, "I have a suggestion. How about if we take the guitar to the gate and ask them if we can take it on board? If not, we'll check it."

Completely disarmed, the guard responded, "Well, I guess that would be all right."

They got the guitar on board and stowed with no problems.

I asked Hanuman what he had taken away with him from that encounter. He said he was amazed that, in the grip of his reactivity, he'd seen absolutely no solutions. And how his girlfriend had seen right through the conflict to a simple answer, coming up with a creative alternative. The critical difference lay in the mode from which each person was operating.

If ten different people confront the same difficulty, you'll see ten different responses. How we react to any given situation depends on our outlooks, our attitudes and assumptions, and our emotional habits—our modes.

Our mode of the moment organizes our entire state of being, shaping what we seek out and notice. Modes dictate our feelings and even what we can most readily bring to mind from memory. Some are toxic ruts; others let us flourish. In either case, they fuel our drives and determine our goals, just as they dictate our moods.

"You can't solve a problem from the state of mind that created the problem," Albert Einstein said. Recognizing when we're in a mode-driven mindset gives us a chance to see more clearly and take the steps needed for change.

Most models in both East and West agree about what our unhealthy modes look like, though they apply differing labels: maladaptive, insecure, distorted, dysfunctional, unwholesome, or even deluded. These unhelpful modes are all contrasted with those that make us more adaptive, secure, wholesome, or wise. Each of these modes, importantly, can move from a negative to a positive way of being.

Our modes of being can be sorted into two main categories: wise or deluded. In our bewildered modes, our perceptions twist in ways that throw us off-kilter emotionally, and we focus on our small-minded world. Such modes bias how we see the world and limit our decisions. But in our wiser modes, we see clearly—without distorting lenses—which spontaneously enhances our empathy.

So for our thinking to be clearer, it is vitally important to learn how to recognize our modes and clarify our perceptions, our feelings, and the actions we take. Our choices can lead us toward the mud of confusion, clouding our perceptions, and toward the murkiness of unknowing; or they can lead us toward the lotus, with its petals unfolding in the light of wisdom, awakening and blossoming, connecting with the nutrients within the mud that allow the lotus to bloom. Rather than succumbing to confusion, mired in the murkiness of blurred perceptions, there's always the opportunity to turn toward *sanje,* the awakened quality that allows the lotus of wisdom to blossom through the mud of confusion.

Buddhist psychology sometimes uses the term "bewilderment" to refer to a haze of confusion in the mind (another term used is "delusion;" Western models speak of "cognitive distortions"). Once we correct the underlying misperception, we can gradually allow a clearer awareness to be revealed.

The Neuroscience of Habit

East and West psychologies agree: a dysfunctional mode does not doom us. These are learned habits; so with reparative learning, we can alter them.

The neuroscience of habit formation and change tells us habits arise because the brain needs to conserve energy.[1] When we first learn any of the endless habitual routines that get us through each day, the brain pays a lot of attention and exerts a lot of energy. But the more we repeat these routines, the less energy and attention is required.

As we practice a routine to the point of mastery, its execution transfers from the higher, conscious part of the brain to the basal ganglia near the brain's bottom. This golf-ball-size brain network guides us as we transfer toothpaste to our toothbrush or change lanes on the freeway while we think about other things—a sign that the brain not only uses little energy for habits, but also that they operate outside our awareness.

The advantage of habits, of course, is that we don't have to think about them as they guide us through our days. The downside is the same: we don't realize how habitual routines lull us into complacency, into going through the same motions over and over mindlessly. While it's great to be able to type without a second thought of where the Z key might be, when those habits are the recipe for our negative modes, complacency hurts. For one, every time we act on these habits we give them more power, actually strengthening the brain's circuitry for them.

Our modes are complex sets of unconscious habits, the result of countless choices that we made sometime in the past and have long forgotten. Our habits may seem the result of thoughtful reasoning, but in truth were established by forces in the mind that we do not notice, let alone understand.

Mindfulness can bring into our awareness workings of the mind that ordinarily are unconscious. When it comes to changing the bewilderment of blind habit into a wakefulness that gives us choice once again, mindful awareness is the key. Recognizing modes in our minds, in our relationships, and in our lives takes our full attention. Lacking that, we are helpless to detect the emotional habits that work invisibly to wield their power over us.

So the first step in changing our mode habits requires that we bring those habits into our awareness, which leads to what I call mindful habit change. The fundamental change that mindfulness brings is one of waking up rather than of being lulled by habitual complacency.

It's a bit like that scene in *The Wizard of Oz*, when everyone trembles as a mighty voice booms, "I am Oz!" The little dog, Toto, goes over and pulls back a curtain, revealing an old man stooped over a control panel and saying into his microphone, "Pay no attention to the man behind the curtain." Our modes have an Oz-like power over us, one that deflates the moment we bring a bold and honest introspection to bear. That true seeing disempowers the invisible grip of mode habits, restoring choice. The clear discernment of mindfulness is like an inner Toto!

Through increasing our awareness of our habitual mode responses and educating the emotions that drive them, the choices we make become less automatic and unthinking, less driven by sheer habit, and more adaptive. Plus, each choice is made with discerning awareness.

Phase Transitions

A friend told me about a time earlier in his life when he was very cynical and negative. He was severely depressed, to the point—though he never tried to act on it—of focusing on thoughts of suicide. His life was falling apart; his wife had left him, taking their young daughter. He felt there was no reason to go on.

An environmentalist by nature, during this gloomy time he went with some friends on a long camping trip to commune with the wilderness of Washington State. On that journey his bitter mode completely shattered and was replaced by a far more positive mode.

One day he was at the top of a waterfall looking out over a gorge at a great expanse of natural beauty. Feeling exuberant, he decided to climb a bit further down the hill while holding on to a large root sticking out of the ground, but the root snapped and he went tumbling. His body rolled over and over down the hill toward the edge of the cliff above the waterfall. His friends watched his tumble in horror, seeing that in just a few more feet he would be gone. But somehow something on the hillside snagged him and kept him from rolling over the cliff.

Perfectly conscious, with only a few scrapes, he jumped up and shouted to his friends, "I'm okay!"

"From that moment on," he told me, "I never had a suicidal thought. Everything felt like a gift." He said that during those terrifying moments he thought mostly about the love he felt for and from the people in his life—and that was the real gift. In the face of death, his mind shifted into a positive mode where he felt filled with love.

Amazed by this miraculous turn of events, I asked him if there was anything else he had learned from this experience that had been life-changing. He reflected for a moment, then said, "I clearly see how the choices we make—motivated from the place we're in at a particular moment—can turn us toward the negative or the positive."

In the years since, he said, he has rarely fallen back into his cynicism, anger, and negativity—at least not for long. When once he had been motivated by anger, he was now more interested in positive solutions. Where once he had seen only disconnection and hopelessness, he now found connection and possibility.

Negative thinking originates in a self-centered focus on what's wrong. The more we focus there, the more we suffer. But if—like my friend after his brush with death—we shift from "my pain, my desires, my attachments" as a central focus, the habit of self-referencing starts to fall away.

We all have a handful of favored modes we enter at one time or another. Their transitory nature allows for growth and change if we spend more time in adaptive modes and visit our unhelpful ones less and less.

No matter how firmly a negative mode grips us or how unhappy it makes us, we always have the potential to shift into a better one. Physicists call such a change in state a "phase transition." Phase transitions occur everywhere in the world of matter, as well as in our minds. Heat a wet glob of clay sufficiently and it hardens into a brick. Heat sand along with a dash of chemicals at high temperatures and behold: glass.

Like water changing to ice or steam, our minds can transmute modes. A confused and agitated mode, when the right interventions are applied, can morph into one of discerning calm and clarity: pick up a baby having a tantrum, hold her lovingly while singing softly, and she just might fall asleep in your arms.

On any journey—particularly an inner one—it helps to have a map. Ours starts with Buddhist psychology's modes of desire, aversion, and bewilderment, each of which can be transformed into a positive set of qualities, such as clarity and equanimity. Developmental psychology looks at the maladaptive modes that arise in our relationships, such as the don't-get-too-close stance of the avoidant mode or the constant worrying of the anxious mode.

Evolutionary psychology looks at modes that helped us survive in the wild: predator and prey. In modern life, subtle versions of these modes play out in a dance of control and resistance. A branch of cognitive therapy tracks a range of modes typified by destructive emotional habits—for instance, the extremely avoidant mode in which we numb ourselves to avoid disturbing feelings.

The secure, or integrated, mode fosters a range of positive, adaptive ways of being in which we seem to flow through life and blossom in our relationships, health, productivity, and creativity. The secure mode is pivotal in this progression, an emotional safe harbor at the core of mental wellbeing, which fosters wiser choices in our lives and stronger bonds in our relationships.

Eastern psychology details a mode range beyond such day-to-day wellbeing: with a wise heart, we undergo a transformative inner evolution toward compassion, wisdom, and equanimity, where life's hassles can leave us unperturbed. Most spiritual paths aim toward some version of this mode.

The spectrum of modes resembles a ladder of inner phase transitions, from heavy to light. At the lower rungs are our more distressing and self-defeating ways of being; like the atoms in steam, our thoughts can be chaotic and our feelings turbulent. As we shift into more healthy modes, we calm down and gain clarity. And the positive range of modes goes into states of mind where our lightness of being transcends the weight of our ordinary ways of being.

Wherever we tend to be on this ladder, we would do well to remember what the Zen master Suzuki Roshi wryly observed: We are perfect just as we are—and we could still use a little improvement.

The modes arrange themselves in a natural progression from most troubling to most liberating. Mind whispering offers tools for psychological phase transitions: ways to free ourselves from the grip of maladaptive modes. On this inner journey we connect with capacities that allow a lotus-like discernment to rise from our bewilderment and confusion, the negative mud of the mind.

Perceiving Anew

When seeing Monet's paintings, we may marvel at how an innovative way of seeing can be catalyzed by a single gifted person inviting us along to see things in a fresh manner.

Art historians tell us Monet's remarkable Impressionist works were partly an expression of the distorted perceptions induced by cataracts. As his vision became increasingly blurred, he went on painting through the years, depicting what he saw.

Monet took great care in his study of the nature of light and the ways it subtly alters what we see. The point for him seems not to have been whether the landscape was foggy or clear but rather the particulars of light that could be captured on the canvas. Monet's fuzzy vision shifted our own from the specifics of objects to the qualities of light cast on them.

Can we allow ourselves to perceive anew, even when our vision is blurred? Can we find wisdom—or at least clarity—in the thick of confusion? Can we miss significant meanings amidst the mind's muddy waters? And could re-perceiving help clear the haze?

A children's book about the senses has inserts to touch, see, smell, feel, and taste, each instructing kids on how we perceive. One of the inserts is a card with transparent strips of colored plastic. The book tells the child to look at an object through one of the colors for ten seconds, and then take the strip away and see how that object looks now.

For a brief period there's an afterimage, a haze of the color complementary to the one you were looking through; if you used the green, you see the world as if through a rose-colored lens, with a reddish patina. But the colored afterglow eventually fades, giving you the ability to see things clearly again—that is, if you don't keep gazing though the colored lens.

How are our perceptions affected by the lenses we see through? What are the lenses we put on time and again? And how can we learn to see our world clearly, free of the bias of our lenses?

Modes color our perceptions. They define how we perceive, and that defines our world. It's not so much the ways life tests us that define us, but our outlooks as we meet those tests. "We are disturbed not by things," Epictetus, a Greek philosopher, observed, "but by the view we take of them."

Aaron Beck puts it a bit differently: "The question comes down to where we focus our attention. If you focus on the negative, that's all you're going to see. If you focus on the positive, you see things very differently." Those who habitually see the glass as half empty are not just more pessimistic; as Dr. Beck has found, they actually are more susceptible to a depressed mode.

Our modes are like obscuring veils. You see a veil as it is, but you also see the world through it. As you become more settled and present, the veils become more and more transparent, like the mud that settles to the bottom of a pond, revealing the crystal clarity of the water above. Whispering, yielding, listening, you can allow the light of awareness that is always there—though temporarily obscured—to illuminate the mind.

Recognizing Our Modes

To begin to understand the unique sensory universe of a horse, put your open hands together in front of your face so that you can't see straight ahead but only to the sides. "That's how a horse sees the world," Bob Sadowski explains. "Each eye talks to only one half of the horse's brain. Watch a horse approach something it's trying to understand, and you'll see that the horse turns its head from side to side so that both eyes can take in the image."

Because their eyes are on the sides of their heads, horses have a huge blind spot directly in front of them. But we humans are used to eyes oriented in front, and we act as though horses are like us—we come at them head-on, or we bring our hands or a halter up from below, another horse blind spot. This can be unnerving or even threatening to a horse. It's even seen as predatory.

So instead, Bob approaches a horse from the side, slowly, the way other horses do. He shows a horse its halter, putting it within sight to the side of the horse's head, and then gently rubs the horse with it, so the horse feels safe. Contrast that with the typical way people approach a horse and put on its halter, and you see that things have been off-track between humans and horses for a long time.

Recognizing our modes is a bit like getting to know how another species, like a horse, experiences the world. Every mode has its own reality. We can begin to recognize when the distinctive signs of a mode arise in our bodies and minds—a flash of anger, the clench of fear, the numbness of avoidance. We might not even realize what mode we are in, but we feel a negativity coming over us, an all-too-familiar knot of unpleasant feeling or a mental fog distracting us from the present.

While we're caught up by a mode, we see our world through those lenses, without realizing that we're seeing through them. But there are some telltale signs that we're in the grip of a maladaptive mode. Our thoughts tend to follow a rut, and our feelings are often way out of proportion. Or our responses are knee-jerk rather than well-considered choices.

What we pay attention to defines our subjective reality at every moment; the more adaptable our attention, the more aspects of reality we can consider. On the other hand, the more we freeze our focus, the smaller our range of choice, and we become self-absorbed.

Three elements of self-absorption all operate while we are in the grip of an unhealthy mode.[2] First, attention fixates on thoughts and feelings typical of the mode itself; we have little or no ability to see other than through the mode's lens, and we ignore what does not fit that view of the world.

Second, we skew the meaning of events (even irrelevant ones) by interpreting them as references to ourselves. We are trapped by egocentric thinking and create distorted personalized interpretations that exaggerate how much an event actually has to do with us.

Finally, we cling to the goals and desires the mode imposes for us, sometimes even at the expense of the wellbeing and rights of other people—and, sometimes, at our own expense.

Getting to know the signs that a negative mode has begun is the first step in changing that set of habits toward a better direction. Once we recognize the mode, we can remember that these are like changing weather patterns, passing states of mind.

While a group of my workshop students studied this together for a year, one of them realized how her lifelong history of repeated relationship difficulties signaled a mode pattern that she had not until then recognized. As she put it, "I see how I have distorted perception. I never thought of it so much as a warped view. I just thought of it as me!"

Acceptance

A man in his late seventies said to me that as he looked back on his life, one thing that he'd found important was "being seen." An attunement to the unique reality of another person lies at the heart of compassion, and a first step in compassion is being able to truly see and understand people, so you can help them effectively.

Life is full of moments when we can practice compassion, if we're attuned to these opportunities. Sometimes it's quite subtle; it may be simply sitting in silence with someone who is going through a difficulty and being available when needed. It may be giving our full attention. Or connecting in a way that allows difficult feelings to rise to the surface, and perhaps even melt in our loving presence.

It's the same when we approach our modes: we need to attune to the unique reality of the mode and understand how it leads us to think and see the world. Attunement lets us sense whatever reactions we are having in the moment; it's at the heart of whispering.

Listening and empathizing help attune us to the genuine essence of people. Once we know that negative modes are part of everyone's packaging, with compassion we can give a nod of acknowledgment yet not get distracted or defined by another person's modes. This applies to our own modes as well.

We each tend to gravitate toward favored modes that have become habitual—they feel like "me." That's one challenge in changing our habitual modes: it can seem as though these habits are just the way life is. If we do not see that we are stuck in a problematic way of being, then we're unlikely to consider changing things.

In all aspects of mind whispering, practicing lovingkindness and compassion toward ourselves and others opens our hearts and counters harsh judgments. In investigating our modes, for instance, we might gently nod to how our toxic modes originated as adaptations to difficult situations and acknowledge that these early survival mechanisms have persisted past their time.

Even in the midst of our mode blowups and meltdowns we can tune in with empathy for ourselves and with compassion for the mode's perspective. Empathy also helps us be forgiving with ourselves about lapses rather than being harshly judgmental or feeling dispirited and giving up.

Empathy and acceptance are vital in this mode alchemy. When in the grip of a distorted mode, it's as though we're the little kid who's been frozen at an early developmental stage. To melt this frozen state, we need warmhearted understanding.

Compassion melts inner barriers while insight releases them, as we integrate fragments of our beings into a larger dimension of our nature. "Just as the caterpillar thought the world was over," as a saying goes, "it became a butterfly."

Root Causes

Here's what I saw the other day when I found myself flipping through TV channels while working out on a treadmill: The animal channel was showing predators, featuring brutal scenes of killer whales devouring seals. The next channel had a reality cop show about a vice squad in Hollywood; its sordid details included a sad prostitute describing how she was sexually abused in her early teens. The third, a shopping channel, offered a seemingly endless series of mind-numbing displays of fire sale–priced zirconium jewelry.

Much of what passes for entertainment can be seen through the lens of Buddhist psychology in terms of the "three roots": attachment, aversion, and their underlying cause, bewilderment (sometimes called greed, anger, or ignorance).

If we prefer one of these modes of relating to the world, it can become habitual, solidifying into a personal style; for example, greedy or angry reactions can become basic stances in life, favored reactions to whatever we perceive.

The roots, in their extreme forms, act as toxins in the mind, each in a unique way. Attachment ranges from having a subtle preference to an outright desire, or from clinging to an intense craving. Stinginess, a refusal to let go, is a coarse form of attachment.

Aversion begins with a subtle movement away, and can become anger or strong hatred.

These modes can also create a mental fog that distorts how we appraise the very object of our greed or dread. That fog is a sign of what Buddhist psychology calls "delusion" or "bewilderment"—the root cause of destructive modes of being.

The Pali word *dukkha*, often translated as "suffering," can also be rendered as "reactivity." The roots can give rise to a wide variety of reactive states, such as worry, envy, agitation, greed, and hatred, as well as distorted perceptions due to self-centeredness, dullness, or closed-mindedness.

At their most extreme, the three roots can create pathology: attachment can become addiction; aversion can turn into hatred, rage, and violence, the drivers of exploitation and racism; bewilderment can become indifference to injustice and cruelty. These are destructive whether at the individual, family, or global levels.

The original Buddhist sources speak of these ways of being as "temperaments." But these modes aren't necessarily permanent; they can be transformed. And any of us can enter these ways of being from time to time, given the right conditions, even if we tend to favor one mode over another.

Our simple attachments and aversions (*I'll have a skinny chai latte with extra foam, please*) guide us through the thickets of life without our having to give it a second thought. But when they go into overdrive, they become what Buddhism sees as mental "poisons," morphing into all-consuming modes of being that twist our perceptions of reality. Then they exaggerate out of all proportion the positive or negative qualities of what we perceive—or befog our abilities to discern clearly—and can destroy our inner equilibrium.

Our problems begin when we don't notice the distortions hidden in our mode thinking or how these distortions drive what we do. Then we view our world through the haze of our attachments or aversions. Whatever happens to us in the moment, these unconscious habits determine our reactions. If pleasant, we habitually gravitate toward; if unpleasant, we push away.

Consider what happened when a couple I know was looking for a new house, and each had strikingly different reactions to the home they finally bought.

A realtor had brought the wife to see a simple A-frame set on a hillside with acres of forest behind. She immediately found it enchanting. There were many features she loved—the sylvan setting, the view over woods and meadows, the cathedral ceilings—but most of all she loved the possibilities. She was put off by the narrow, small rooms and the "tacky" quality of materials, but her inner decorator immediately set to work imagining how they could remodel some rooms and how different colors, tiles, and upgraded materials could transform the house into something they could love.

Her husband, on the other hand, hated the place at first sight. The house had been built on the cheap by a shop teacher, who had apparently deployed a high school woodworking class to construct it. Some of the rooms were tiny, and there weren't always right angles where walls met. He fixated on the bargain materials that had been used, like the boxcar-sale vinyl-clad plywood paneling that covered the walls. He loathed the black plastic "raw timber" that imitated wooden beams on the ceiling. The only thing he liked about the place was a lovely freshwater spring that emerged from the hill in the woods behind the house.

When they brought a friend over to give her opinion, she didn't help much. She seemed oblivious both to the house's potential as well as to its downsides. She barely registered the fact that they were eager for her opinion of the place. She deflected their questions about what she thought with a vague, "I don't know." Then she fell into gossiping about some acquaintances and deciding which restaurant they should try for lunch.

Those reactions each exemplify modes based on the three roots. The wife's positive take on the house, focusing on what was attractive and how it could be improved, reflects the attachment mode. The more she thought about the house, the more she wanted it. In its extreme, this mode fosters clinging and craving.

Her husband's negativity typifies the aversive mode, driven by a knee-jerk disliking. In this mode we fixate on what's wrong rather than what's right and dismiss other possibilities; in more intense forms, this mode can become not just rejecting but hostile.

And their friend's obliviousness to the whole situation is typical of the deluded, or bewildered, mode, where confusion and lack of focus create a chronic indecisiveness or indifference; the extremes here can go beyond cluelessness to a dull ignorance.

Noticing Clues

Back in the fifth century an astute Indian text described how something as simple as how a person sweeps a room offers a clue as to which mode that person prefers.[1] In the attachment, or "greed," mode, people sweep gracefully and move with composure, elegance, and a spring in their steps. When they enter a new place, they notice immediately what's most pleasing there. Likewise, when they meet someone new, they are most struck by the positives in their first impressions, sometimes failing to notice even glaring faults.

This attachment mode takes grasping and sense pleasures as its organizing energy. In this mode we want more and more, cling to comforts, and avoid discomforts and disharmony of all kinds. Among the unhelpful states of mind frequent in this mode are avarice, self-centered willfulness, vanity, pride, jealousy, and deceitfulness.

In contrast, people in the aversive mode sweep with harsh, almost angry movements and more generally relate to the world through a negative lens, which readily rejects what it sees and makes a steady stream of criticisms. They notice what is wrong, not what is right. A fellow who could be a textbook case of the aversive mode once confided, "My first reaction to anything new is to say no."

Through the aversive mode's lens, people see faults and problems wherever they look. Their judgmental appraisals make them readily displeased, disparaging, and quarrelsome. When this mode has a strong hold, people are prone to being tightfisted, rigid, vindictive, even cruel, and sometimes haughty. Those in the aversive mode move through the world in a rush, short-tempered and impatient, with their bodies tense and tight.

At an extreme of the aversive mode, people express hostility in the raw, impulsive, uncensored way a child does—for instance, blurting, "I hate you!" In a more sophisticated, grown-up version of this mode, people express their aversion more subtly, in hurtful critiques, sarcasm, or witty put-downs.

Then there's the deluded, or bewildered, mode, where confusion marks a person's perceptions and appraisals, and perplexed agitation and worry result. What might be called the "willingness to know" gets temporarily shut down. Those in this mode are indecisive, often not quite knowing what to do, and they readily resolve this discomfort by ignoring what's happening. The deluded mode's scattered thinking and indifference make it ripe for spacing out.

On the sweeping test, people in the bewildered mode are haphazard and sloppy. They arrange their things in a chaotic, messy way. They move with muddled hesitation, walking with an uneven shuffle. In social settings they just follow the lead of others. That ancient text even describes sleep in this mode: people sprawl facedown and wake up dazed.

Jack Kornfield, a meditation teacher who has written about Buddhist psychology, gave a seminar on these three mental tendencies. He had people gather in affinity groups to discuss what their mode was like and present it to the rest of the participants.[2]

As the greed group gave examples of ways they tried to manipulate circumstances to get the best outcome for themselves, they were polite and worked in harmony. The aversives came up with tales of judgments and interpersonal problems, and predictably got into disagreements with each other—even with the people to whom they were presenting. And the deluded group's presentation was unfocused and confusing.

To sum up each of these modes in a question, the greed mode asks, *What's the best?* The aversive, *What's wrong?* And the deluded, *What's up?* (Or maybe just *Huh?*)

Approach or Avoid?

Once when I was visiting the British Virgin Islands I lost my comb. After a few days of going around with an ever-wilder mane, I went to the local drugstore to search through their collection of combs.

There seemed to be a wide selection for all kinds of hair—wide-tooth combs, small combs, and all in an assortment of colors. They had just about every color except my favorites. I was searching for a red or a black. I'd even have settled for a white one. But green or yellow—I'd rather grow dreadlocks!

The seemingly bemused West Indian storekeeper was watching me sort through all the combs. She came over with a tolerant smile and asked with a chuckle, "You color choosey?"

At that, I quickly settled for a brown comb and playfully pretended to neaten my frowsy locks.

Ask any kid what his favorite color is, and you'll get a ready answer—our rudimentary sense of self gets established in the second year of life, with all our preferences and self-interest growing from there. The ease with which we register such a simple like or dislike reflects a basic design in the brain that embeds this choice in its two halves.

In all species, from reptiles on up, the brain's right side has a dominant role in avoiding, the left side in approaching. In most people it's the right side of our brains that activates more when we find something disgusting or threatening and the left when we desire or like it.[3] Everything we encounter gets instantly categorized by the amygdala circuitry as good or bad, liked or disliked. Kitty? Scorpion? We don't have time to give it a thought—the choice gets made for us instantaneously.

This approach-or-avoid choice is so deeply embedded in the mind's operations that many psychological theories—and most every mode model—reflects it in one way or another.[4] Approach is the core of the attachment mode; avoidance is at the heart of the aversive.

This positive or negative emotional weighing of each thing we perceive sets the stage in the mind for a cascade of appraisals, thoughts, feelings, and actions, which can blossom into a greedy wish to possess the object or get distance from it in disgust, fear, or anger. If an experience is pleasant, we gravitate toward it—the seed of clinging. If unpleasant, the mind glides away, a tendency that can grow into aversion. And if we don't notice that the mind clings or feels averse, this lapse in awareness can be the doorway to ignorance.

Attachment and its turbocharged form, craving, are "approach" emotions, driving us to seek out, grasp, and possess the objects of our wanting. When in this mode, our perceptual bias enhances the positive and ignores the negatives of our objects of desire. The fog of greed blinds us to any less-than-desirable aspects of the very things we long for, whether a new electronic gadget or a pair of shoes.

And when we get what we want, desire does not end; it simply finds a new target. When lost in the mode of greed, we are constantly dissatisfied—like being intensely thirsty and drinking a glass of salty water, which leaves us even thirstier, so the more we drink, the more we want. That kind of thirst never gets quenched beyond a temporary satisfaction, just as that kind of craving never ends. A friend once told me about being addicted to blissful states in meditation!

Like the approach emotions, the avoidance emotions have their own distortions, which obscure the positive qualities and accentuate the negatives of whatever we perceive. We want to distance ourselves, since avoiding what we find so unpleasant can give us temporary relief.

But the underlying aversion does not go away. The unpleasant feelings will come back eventually when that object—be it a thing or a person—reappears, because the same fundamental disturbance of mind remains ready to rise up when conditions are right once again.

On the other hand, desire is not in itself negative. We can have happiness and pleasure without being attached.

These modes, Buddhist psychology holds, can be transformed through the right practices to their positive aspects: attachment becomes a sensitive discrimination; aversion becomes a clear discernment; and bewilderment becomes a spacious equanimity.

The Primal Choice

As I was standing at the kitchen sink looking out on our garden one spring day, I saw a small rabbit standing stock-still, not moving even a whisker. The rabbit had also noticed me through the same window and was riveted. As I had reached for the faucet, the rabbit had frozen, its little nose wiggling from side to side, as though sharpening its senses. Its pointed ears stood straight up at attention and its eyes stared with an unblinking gaze, while it carefully assessed the potential for danger.

After a few moments of heightened scrutiny, the rabbit relaxed its bunny mind, softened its body, and went back to eating the grass.

As a prey animal, rabbits survive by being exquisitely attuned to potential dangers. This rabbit was scrutinizing me to determine whether it was safe from the sudden presence of that unfamiliar shape looming behind the window.

That rabbit wasn't being paranoid but smart, deploying the mode of bunny mind that no doubt helped preserve its ancestors from danger after danger. Just as the human version of bunny mind functions to keep us safe, our most primal modes are legacies of ancient human prehistory, a rugged and dangerous time when these modes were essential survival strategies for our species.

Whether to approach or avoid is even embedded in the genes. If a rodent pup gets lots of licking and grooming from its mother—a sign that there is an adult to protect it from being seen by, say, a hungry hawk—that pup's genes are turned on in ways that make it more readily approach the novel and unknown.[5] But if, sadly, the pup gets no licking and grooming—a sign that its mother may have died and left it vulnerable to hawks or otherwise neglected—its genes flip into an avoidant pattern, which makes it anxious and fearful, afraid of new places and things, averse to risk.

The modern descendants of these primal modes operate in our lives. The exploratory attachment mode lets us discover new vistas and possibilities. The wary aversive mode keeps us safe and out of trouble in the modern version of a jungle: the city.

Each of these modes of operation embodies a skill designed to solve a specific problem in life, and each gets activated when the relevant cues for that situation come our way. We don't have to think about which of these to choose; they are chosen for us automatically, out of awareness—hopefully the right one at the right time.

It all happens in just moments. Such instantaneous switches into the flight mode offer a bunny—or you or me—rapid guidance on what to do to survive. This split-second mobilization tells us what to do far more quickly than if we have had to stop and think over every possibility before taking action.

The brain circuits for negative mode triggering operate via an ancient survival tactic that we carry into everyday life. To ensure we are safe, the amygdala, an emotion center deep in the brain, constantly scans our surroundings for threats. When it spots danger, our amygdala triggers a fight, flight, or freeze response. Even when the threat is just symbolic, our fight-flight-or-freeze response can tumble us into a dysfunctional mode.

The amygdala is part of the vast mental circuitry that works largely outside of our awareness. Mode triggers act behind the scenes and in a sneak attack can take us over with great rapidity, blindsiding us. Our rational mind hasn't a clue that the amygdala circuitry is launching a toxic mode until we find ourselves already plunged into one.

Modes operate along a spectrum of energy levels. Some events might prime a mode, adding a drop of energy to its charge. A subtle cue might only prime the mode a bit, but as we encounter successive or stronger primes, the charge of the mode increases accordingly. As the charge builds, a mode can become stronger and more active—with the potential to dominate our thoughts, feelings, and even our bodies—and shape our inner realities.

Fight, Flight . . . or Freeze

My special affection for rabbits dates from the fourth grade, when my brother and I had two bunnies, Snowflake and Lucky. We treated them like pet dogs; even in our New York City neighborhood they would wander freely around the streets.

Everyone on the block knew them. They were quite friendly, perhaps overly so. One afternoon they had been gone for too long, and my brother and I went looking for them. No one had seen the rabbits.

We finally found them hopping together in the window of the neighborhood butcher shop! Fortunately we rescued them, and they hopped behind us as we led them back home, neighbors cheering as we went.

As Snowflake and Lucky found, life holds real threats. The fight-or-flight mode in the face of an actual danger mobilizes us to take action that can be lifesaving. Just as with that bunny in my garden, scanning to see if there was a real threat, our circumstances determine whether a given mode is useful or not.

Those of us who are too often caught up in the fight-or-flight mode needlessly flood our bodies with stress hormones, preoccupy ourselves with useless worry, and otherwise waste mental and emotional energy. Mobilizing the fight-or-flight mode too often blocks a different way of being that is just as vital: our mode for rest and recovery. Our body needs time in this mode to stay healthy.

The recovery mode puts us in a state driven by the physiology opposite to that for fight-or-flight: the parasympathetic, rest-and-restore system, which allows the body to build capacities for resilience and effectiveness. Our attention flows freely as needed to achieve our goals of the moment.

Such modes are holdovers from long ago but come to us so naturally today because they are imprinted in the basic blueprints for our central nervous system. In addition to the hardwired, built-in modes ingrained over evolutionary time, each of us acquires another set of modes that helps us adapt to the challenges of our early lives—and which we retain as holdovers throughout adulthood.

For someone growing up in a tough neighborhood (or family) where bigger kids are bullies, a highly cautious mode can be adaptive—for that period of life. If that once-adaptive mode persists and dictates our ways of being later in life, when it no longer serves us—if, say, we walk around with a chip on our shoulders—the overreaction can become a problem rather than a solution.

Hyper-cautiousness reflects the primal mode for survival, but today it might mean we misread a neutral face as threatening and therefore have a misunderstanding with a lover or miss an opportunity to connect with a new friend.

To loosen a mode's grip, we need first to realize the ways it is counterproductive—an insight that the mode itself does not allow. While swept up in a maladaptive mode we might even believe we are getting what we want, a misperception that results from how the mode shapes our thinking and our reactions. Modes operate with faulty reasoning, which is harder to correct while we are under a mode's influence.

The crux of the dilemma with our dysfunctional modes is that while we are in their grip, we're often helpless to find ways out of the difficulties they create for us. When toxic modes capture our minds with their distorted lenses, mistaken assumptions, narrowed focus, and faulty reasoning, it's as though we have fallen asleep or are going through parts of our lives in a trance. Choice vanishes.

CHAPTER

Insecure Connections

A surgeon was sued for malpractice. She had endured ten days of a harrowing trial, during most of which she had to listen to terrible things being said about her. She was not permitted to respond to any of it until the very last day of the trial, when she finally had her turn on the witness stand.

When that grueling day ended, she went home, took a hot bath, and got under the covers in her bed. Her husband went out for pizza and then joined her in bed, where they devoured the pizza, talked over what had happened, and waited for a phone call from her lawyer about the verdict. (Happily, the verdict was in her favor.)

"Under the covers," she said later, "was the only place in the world I wanted to be."

The emotional power of such a cocoon, and why we find it so soothing, has its explanation in the pioneering work of British psychoanalyst John Bowlby. It lies in our lifelong need for a sense of security and for people who give us that feeling, just as a mother's reassuring hug calms a child.

As adults, we turn to spouses and partners, friends or therapists to provide the security that we once got (or yearned for) as children.[1] These are the people in our lives who we rely on in moments of upset to help us manage our distress, as that surgeon did with her husband in their cozy nest.

As Bowlby found, the expectations and habits we develop as children are also those we bring to our close relationships later in life, and they are largely molded by our interactions with the key people in our small universe: our parents and families, our friends and teachers—all the people with whom we have repeated interactions. How caregivers meet (or fail to meet) a child's needs for attachment, security, and protection have lifelong consequences.

There are two major insecure paths in life. Children who have caregivers who are cold and distant, or inconsistent—sometimes warm, other times absent, for example—develop core assumptions, or working models, that lead them to be insecure in their relationships as adults.

We bring such deep beliefs to our relationships later in life. Our past becomes the map for our future connections. The more intense and prolonged the experiences that shape the outlooks, the stronger the learned reaction patterns will be.

These insecure styles come in two main forms: the avoidant mode and the anxious mode. Both modes begin as temporarily helpful adaptations to the difficulties of childhood life, when caretakers fail to provide a secure base.

The most influential research on these relationship modes in adult life has been under the leadership of Phillip Shaver, a psychologist at the University of California at Davis.[2] I met Shaver when he had come to Dharamsala, India, to participate in a five-day meeting with the Dalai Lama and a panel of other scientists on neuroplasticity—how repeated experiences reshape our brains.

As I heard Shaver explain these insecure modes (which he calls "attachment styles") during his presentation, I had an epiphany: I recognized patterns that are shared with our more primal template for survival, to approach or avoid, as well as with the Buddhist mode model.

But rather than dealing with a threat to survival, the insecure modes each become a strategy for managing our relationship insecurities and distress. These modes mainly trigger in the relationships that matter most to us, creating puzzling eddies and troubling storms.

In the anxious mode, we try to master our anxieties about connection by incessantly thinking about it, ruminating endlessly. In the avoidant mode, we try to dodge relationship anxiety through suppression, diminishing our passions by cutting ourselves off from emotion. Each pattern, says Bowlby, tends to be self-perpetuating.[3]

To recognize the insecure modes, it helps to start by contrasting them with a quick look at the secure mode. Consider whether these describe you:[4]

- "I'm comfortable being close to people."
- "I don't get anxious when someone close to me goes away."
- "I feel secure in my relationships."

If you agree with those statements, you are likely someone who readily enters the secure mode, typified by a sense of satisfied wellbeing in your relationships. In chapter 7 we'll explore the secure mode as the emotional foundation of a range of adaptive, positive ways of being.

Then there are those of us whose childhoods were not all that secure. One of the insecure modes may have helped us in childhood. The problem comes when these modes persist into adulthood.

The Avoidant Mode

Toward the end of a long flight from San Francisco to Boston an elderly passenger had a mild heart attack. He sat ashen-faced, slumped in his seat, as the flight attendant frantically announced, "There's a medical emergency. Is there a physician on board? Please identify yourself immediately."

Two doctors rushed down the aisle to the ailing man and administered to him as his daughter sat crying in the next seat. Thankfully, they were able to revive the man, who finished the flight wearing an oxygen mask. By the time an ambulance crew met him after the plane had landed, he was in good spirits.

As the rescue had unfolded in midair, the other passengers had sat riveted on the life-and-death drama happening just a few feet away— that is, almost all the passengers. A few didn't look at all; they just continued watching the movie or reading a book, as though nothing untoward was occurring.

Looking away rather than engaging with a disturbing scene typifies the avoidant mode. Those prone to this mindset have learned to avoid confronting upsetting situations in order to help them suppress anxious feelings. Such people typically agree with statements like:[5]

- "I am uncomfortable being close to others and find it difficult to trust them completely or to allow myself to depend on them."
- "I am nervous when anyone gets too close."
- "I don't like to think about whether my partner really loves me or wants to stay with me."

In the avoidant mode, people cope with distressing feelings by dodging them with whatever maneuver keeps the upsetting feeling at a distance—distraction, denial, repression, going numb. This includes actually running away: leaving a relationship that has become troubling or merely threatens to, even if that threat is just imagined. And within the mind this maneuver can mean failing to look squarely at a hard truth.

An extreme of this mode involves detaching from our emotions altogether. While caught in this sort of detachment a person feels that his or her life is empty and boring and withdraws into an aloof and cynical stance or into a private, "spacey" reality, compulsively pursuing distractions that soothe or stimulate. (Pointlessly flipping channels or surfing the web for hours may be signs of this.)[6]

Although those given to the avoidant mode may look perfectly secure to someone else, this is often because they are reluctant to disclose their fears. Or their suppression may be so strong they do not even experience their anxieties.

The avoidant mode's consistent failure to engage feelings can mean that those prone to it lose touch with their own emotions and find it confusing to try to tune in to them, if only from lack of practice.

While in the grip of this mode, people tend to be detached, distant, cold, and stay on the surface in their interactions. In their closest relationships they can seem numb, as if they have abandoned any hope of receiving caring affection. That may be because this mode typically establishes itself when a child's caregivers react to his or her distress by being emotionally cool or distant, or showing outright disapproval, rejection, or anger. The child learns to suppress the urge to reach out and becomes overly self-reliant. Such children get along on their own by learning to avoid feelings of distress.

"My first reaction to anything new or emotional," as a friend prone to this mode said, "is to run away from it." (Sound familiar? The avoidant stance has similarities to the aversive mode's knee-jerk negativity.)

The avoidant mode frequently kicks in as a response to thoughts that activate unpleasant connection needs, memories, or interactions, like times when people disappointed or betrayed us. This includes anything that might threaten our sense of connection—notably rejection, loss, or neglect.

Even warmth can trigger an unconscious expectation of rejection. I once saw a cartoon that captures the avoidant stance: a man and a woman are talking while seated at their dinner table. She says, "I love you." He replies, "Don't threaten me."

The avoidant mode skews our attention away from the positive signals of others; even a partner's loving gestures can be missed or spurned. Or those warm cues register too lightly, so later we have trouble remembering these positives. The tragedy of the avoidant mode is this: the love we yearn for may be there, but we will not let ourselves receive it.

The assumption that we can't have anyone's emotional support leads to a striving for self-reliance—that is, not to need anyone, and even rejecting someone's help when offered. We try to keep our needs for connection suppressed or deactivated, no matter how this may damage our actual relationships. Because we fail to reach out to loved ones for support when we need it the most, we leave them feeling as if we do not need them.

I remember a client who traced his avoidant pattern to his relationship with his emotionally distant, always-busy mother. As a child he had often felt there was some project that was more important to his mother than spending time with him. Years later he realized that he had made a sort of unconscious adaptation, resigning himself to never having a full connection with his mother—an emotional stance that carried over into his love relationships. People close to him would sometimes complain that he—like his mother long before—was distant and emotionally unavailable.

As he put it, "I feel I can't allow myself to enjoy the people I love. I do something to sabotage the closeness or pull away inside."

When his avoidance triggered hurt or anger in the rejected person, he would feel bewildered: clueless as to why they had reacted that way and not knowing what to say or do to repair the breach, nor how to change his knee-jerk reactions for the better.

While the avoidant mode provides a façade of independence and calmness, it hampers our abilities to deal with life's difficulties.[7] And, sadly, the avoidant mode's very strategy for managing emotions in relationships often leads to disconnection. When we close down to avoid the negatives, we also close down to what is positive. We can then miss what *is* working, leaving us unable to receive our partners' love.

The Anxious Mode

MOM SOMETHING BAD HAPPENED CALL ME

That ominous text message came to a business executive while she was out of town running a meeting. The text was from her twenty-year-old daughter, who had suffered kidney failure a few years before and was having difficulty adjusting to a kidney transplant.

The executive's heart started pounding, and she excused herself, went into the restroom, and tried to reach her daughter, to no avail. At that, she burst into tears. Her body started shaking. After several minutes, she was able to compose herself and return to the meeting but, as she put it, "I couldn't focus on anything else. I was brain-dead."

Only at the end of the day was she able to reach her husband, who told her that it was just a problem her daughter was having with the registrar at college and that he had gotten her the information she needed. Everything was okay.

Of course, anyone would have been worried. But when she told me about it at a seminar, the executive said she obsessed about that text from her daughter for days afterward. Also, she said, she finds herself fretting about things in her life that are trivial—not just potential crises.

She over-worries about everything and anything, particularly if it has to do with her connections to the people in her life. As she told me, "I just can't go on like this."

Such continual rumination marks the anxious mode. Here are some attitudes that capture the emotional world of the anxious mode:[8]

- "I want to be very close to my partner, and this sometimes chases them away."
- "I worry about being abandoned."
- "I resent my partner when he or she is away, and I'm frustrated when he or she is not there for me."

One common childhood root of the anxious mode begins with a main caretaker being self-absorbed (as is the case with narcissistic, alcoholic, or workaholic parents) or overly anxious herself, to the point where the child gets little attention.

In response, some children learn that to get the caring attention they crave, they must "protest." They become demanding by crying, having tantrums, or clinging. That may sometimes work, getting them the engagement they need to calm down, or sometimes not. But amplifying their needs then becomes an emotional habit.

Of course anxiety has its proper place in our emotional repertoire; a right-size dose of worry can mobilize us to meet a challenge well. But the tipping point from apt concern to the anxious mode is reached when added anxiety does not lead to constructive action. From that point on, we just stew in our worries. People caught in the anxious mode can feel overwhelmed, needy, fragile, and emotionally helpless, or have deep self-doubt.[9]

In the anxious mode, we are vigilant for threats, talk incessantly about our doubts, fears, and needs, and fret, for instance, about whether the important people in our lives care about us enough to respond to our needs. Such ruminations lead us to anxious clinging, dependency, and a craving to maintain contact.

When a relationship gets rocky, this mode makes us hang on for dear life, no matter what. People prone to the anxious mode are more likely to stay in a failing relationship long after its rewards have waned.

Worries, negative emotions, and anxious preoccupation are, oddly, essential parts of this mode's desperate strategy for connection. Mark Twain captured this tendency to exaggerate problems: "I am an old man and have known a great many troubles. But most of them never happened."

These turbulent feelings fit with the unfulfilled wish to make people pay more attention and be more nurturing and emotionally available. And so this mode keeps these emotions in play, even intensifying them, in a bid for the connection it craves.

That seems to be why people in an anxious mode sometimes exaggerate their own helplessness and amp up the seriousness of their troubles. One obvious downside: this urgent search for emotional rescue can, paradoxically, interfere with actually solving the problem at hand, because a solution would thwart having someone come to the rescue.

The anxious mode can foster an unconscious desire for perpetual problems to justify this continual cry for help. Actually solving the problems becomes irrelevant—what's really wanted is someone to come to the rescue, emotionally.

There are several ways the anxious mode sustains its state of hyperalarm. One is to exaggerate the threat by a perceptual twist that highlights the dangerous aspects of even benign occurrences, emphasizing how things could go wrong, all the while seeing events as beyond one's abilities to handle.

When their demands for attention and love fail, people in anxious mode search for ways they may have acted that have lost them their partners' attention, while also being angry at the partners for not caring enough.

The tragedy of the anxious mode comes down to this: despite a history of frustration with their relationships, these people still harbor the hope that they can get the love they want if they amplify their worries, overreacting with concerns about disconnection and demands for attention and closeness.

Modes in the Brain

Vividly imagine that you and your partner are about to break up or that you are about to lose a close friend. Such scenarios can trigger our favored insecure mode, if we're prone to one.

This is the reason Phillip Shaver's research group had young women who had volunteered for a brain-imaging study visualize just such a breakup in their romantic relationships.[10] As the women let their fears about losing their partners run wild, a brain circuit lit up, one that operates specifically when we worry about relationships but not about other anxieties. As the women imagined this upsetting scenario, the imaging revealed clear differences in brain activity between the avoidant and the anxious modes.

Those in the anxious mode showed heightened activation in what might be called the "what-if" cortex, a neural zone linking the emotional centers to the circuitry for thought, which fires up when we worry. This brain area fuels a self-amplifying distress cycle of rumination, which keeps going even after the threatening situation has long been over, imagining how the situation might have been handled differently or mulling insoluble questions, like what this or that person might have been thinking.

Women prone to the anxious mode were unable to shut down their worry circuitry, especially if they were preoccupied by worries about their romantic relationships ending, but they could easily turn off that circuitry when trying to quell anxieties about other parts of their lives.

By contrast, women prone to the avoidant mode showed a very different neural pattern. When these women were asked to squelch worrisome thoughts about their relationships breaking up, a brain area that suppresses upsetting thoughts became active.[11]

Those in the avoidant mode were unique in using this particular circuitry to quiet their worries. They also had steady activity in an area which spurs disengagement, sadness, and passivity. And when the avoidant women were asked to imagine a sad breakup and then to turn those thoughts off, they had more trouble feeling sadness than they did suppressing their worry. Indeed, just as brain activity in the anxious mode can't stop worry-generating circuits, those prone to the avoidant mode seem unable to stop their *suppression* of distress.

Again, as with the modes of clinging and aversion, we can see echoes of our evolutionary legacy and its strategies to approach or avoid: the anxious mode can't stop engaging with troubled feelings; the avoidant mode won't go near them.[12]

This research suggests modes have a signature pattern of brain activity (though there is a lot yet to discover about this). If so, as we switch modes, we are leaving one pattern of brain activity and shifting into another.

An Evolutionary Arms Race

Eat or be eaten? This is the single most crucial decision any animal makes, one we share with even worms and amoebae.[1] Over the millions of years our ancestors' brains were being shaped, we were both predators to those species we could eat and prey to those who dined on us.

The evolutionary arms race between predators and prey has made these among the most primal modes in the human repertoire. "To know ourselves, we have to know our own animal nature first," says Ian Mc-Collum, a Jungian analyst. "The game we are playing is a shared one. It's called survival. The psychological instincts of the predator are in our history and in our blood."[2]

In thinking about how we humans manifest our predator-like and prey-like modes, we need to separate them from their original function during the hunting and gathering years of our species. Today they surface in far more subtle ways: as thoughtless entitlement on the one hand or self-sacrifice on the other.

The predator-like mode, in the sense I mean it, can manifest as being controlling, entitled, lacking in empathy in varying degrees, and even being manipulative (and so the mode can be thought of as predatory control, or predatory entitlement, and so on). These are destructive qualities in any relationship, whether in partnerships between people or between nations.

How often do people approach each other with hidden agendas, preferences, or wants, without tuning in to what's going on for the other person? This controlling aspect mode gives us a blind spot for another person's needs in the moment. As Bob says, "Don't put your purpose before your connection."

Once you learn about the subtleties of such predator-like behavior, you may recognize it in all kinds of interactions. For instance, some salespeople treat customers this way, pouncing on them the moment they enter a store and trying to coerce them to buy something. This alienates customers, who are then less likely to return.

Research on sales stars shows that they start with empathy—the ingredient missing in this mode. They first try to understand what the customer wants, and then act on their behalf. They discourage their buying the wrong thing even if it means losing a sale, thereby winning their trust, loyalty, and repeat business.

There are people who act in such mildly predator-like ways all too often. They try to get their way, to dominate. In this mode, people are aggressive, self-interested, and ready to take over. They can be effective in some ways—taking bold risks, getting things done—but at an emotional cost to those they push around.

Leaders in this mode tend to be clear about what they want and confident in moving toward their goals; they can have charismatic qualities. But when such narcissistic leaders lack genuine empathy for those they lead—and everyone else—they often end up torpedoing their own careers. The potential upside of the mode comes when empathy gets added; then a leader can have clarity and confidence without resorting to a command-and-control style.

Entitlement can be another aspect of the mode, when someone projects an arrogant image of superiority when he actually feels flawed underneath. Narcissists often have this dynamic and are oblivious to the needs of those around them.

On the other hand, the human expression of the prey-like mode leaves us passive, doing what other people want us to, as though we were helpless. This mode often means surrendering to someone else's predator-like mode in order to preserve the connection, or from a self-sacrificing habit where we are out of touch with our own needs.

That connection is nothing like joining up. It can come at the cost of letting the other person control or neglect us. But when someone in this mode engages with people who are not controlling, it can show its upside: attuned cooperation and collaboration, the human version of herd dynamics.

For evolutionary psychologists, much of what we think of as human nature can be reduced to the modes, or programs, we all have hardwired in our brains. Our minds come pre-packed with modes of operating that represent deep habits engineered to solve the problems early humans faced. Each was essential in the ancient past, during the epochs our earliest ancestors survived by hunting and gathering, but may be irrelevant or maladaptive for life today.

These primal modes were not just for challenges, such as escaping predators or catching or gathering food, but also for needs, such as fighting or creating harmony, falling in love, raising children, and protecting our loved ones and us. The operating modes that were successful then live on in us today as primal, automatic programs for engaging life.

This legacy makes our minds something like crowded zoos, with ferocious hunters, frightened prey, cunning commanders, brave leaders, eager lovers, caring parents, and many more vying to take the helm and pilot us through one or another challenge.

The Herd Hierarchy

The art of whispering has revolutionized how people relate to horses (at least in the Western world). For centuries people have used sheer force to break a horse's will. In the past few decades, though, horse whispering has begun to change that radically by treating a horse as a full partner in cooperation.

I was working with Sandhi and Bob, and had decided to bring Sandhi to another field, a meadow with high grass that she loves but was rarely allowed on because she might overeat and founder.

So I started walking off, leading her with a rope. But Bob stopped me, saying, "You can't just walk off like that. You just abandoned her!"

Sandhi had been completely present with me when I suddenly cut off my attention to her. Bob added, "You've got to tell her what you're doing! Connect with her first, bring closure to what you've done, and then tell her what your intentions are."

He suggested I reconnect by patting her and telling her what she had done really well that day, and how we're going to another field now to try something else. When I turned back and stroked her mane and told her what we were going to do next, I could feel her heart melt. She nuzzled me, welcoming me back to our connection.

When I had just walked away from Sandhi, I had begun to enter the predator–prey mode most typical of how humans have related to horses and other domestic animals—not to mention other people—over the eons. In that mode, we enter into the ancient dance where one species dominates another by imposing its own agenda.

Prey animals don't need to be forced to learn; they're willing to collaborate. They react to forceful control as being predatory, and it can make them resist or want to run. We humans tend to assume they are like us, that they perceive as predators do and so need to be controlled or given punishment when they don't comply with rules. But if you understand how they perceive, they are happy to comply, without the need for force.

"Taking control of a horse against its will is predatory," said Bob. "But if the horse trusts you then it will accept you as its leader. Prey animals are relieved when they know they have a leader they can trust."

When you are joined up, the horse collaborates with you and agrees to learn what you're trying to teach. You do not need to force it. It's not about being predatory, not about controlling horses, but about learning together and allowing a horse to trust you as its leader.

After horses learn something, Bob says, they need a little time to chew on it and take it all in. When they retreat, we need to respect their "bubble time." As Bob advises, "You need to recognize when to retreat; it sets up a connection."

And always we need to attune to the horse's rhythm and join up. "Move slowly, like a prey animal," Bob advises. "Predators move fast. Horses aren't afraid of predators but of predatory behavior."

Horses, like many other prey species, have survived by learning to be ever vigilant for predators and nimble in escape. Horses live with a sense of relatedness. As prey animals, they rely on and are highly attuned to each other, joined together in their herd dynamics. While horses graze they often take positions that let them scan a horizon that the other horses don't see. As a group they create what amounts to a circle of vigilance.

Humans, in the eyes of a horse, are typically viewed as a form of predator. When we treat a horse—or a person—as though it were simply an object to manipulate, we confirm that view.

Nature made humans capable of being both predator-like and prey-like. (Perhaps that explains why we're a bit confused.) Today this means we are potentially capable of playing either role in our predisposition toward others.

Bob pointed out that the predator–prey dimension interacts with another: dominance and submission. Animals that live in groups establish a pecking order. Some are in control and others submit to that control. This can be quite benign, as with a mother being protective of her young child, or malevolent, as when a mean-spirited dictator controls an entire populace at his whim.

"There's a natural herd hierarchy among prey animals that can be seen in how they take space," Bob said. "A dominant horse may want the grass you are on or want to move you to another field."

"That looks predatory to me," I challenged him.

"Predators stalk their prey and plan their attacks in a premeditated way," Bob replied. "Horses don't do that. For a horse it's more about having a clear herd hierarchy and a leader who provides safety for the herd—it's a herd partnership."

He added, "But with predatory dominance, you see complete control. Predators have an organized, premeditated plan of attack based on what they want."

Then I raised the question of "enabling" predatory behavior in human relationships. I was thinking, for example, of the dynamic between an entitled narcissist and someone who is resigned to complying with the narcissist's demands. "Is extreme prey-like behavior too passive?"

Bob thought about it, then said, "If a prey animal's chance for 'flight' is taken away from it—if you cut off its escape routes—it does fight, if it needs to protect itself, even though it would rather run."

This distinction between prey and predator brings new insight to issues of control and passivity in human relationships. Prey dominance acts as a corrective to predatory dominance, just as being assertive corrects passivity on the one hand and aggression on the other.

For humans, I think of prey-like dominance as assertiveness, which can include tough love. If predator–prey is our only model of intergroup relationships, the outcome is an evolutionary arms race in which two groups must continuously strive to outdo the other. Gazelles run fast, so cheetahs run faster; similarly the Cold War arms race led to "mutually assured destruction," where if one side started a war, both sides would be destroyed.

This raises the question, when do modes go beyond their functionality, becoming outdated relics of our past? When is it time to move beyond the survival-of-the-fittest mode to one that allows for mutual cooperation, tolerance, and respect?

Evolution offers a peaceable alternative: co-evolution, where two species influence the development and survival of the other. Flowers, for instance, owe their beauty to the insects that pollinate them. A variety called Darwin's orchid, found in Madagascar, has a narrow, eighteen-inch-long tube that leads to its nectar. When Charles Darwin heard about it, he proposed that there would have to be a flying insect with a tongue that long to pollinate the orchid. And a few decades after Darwin died, a moth was discovered on Madagascar that specialized in pollinating just that orchid—and its tongue was eighteen inches long!

Empathic Pauses

The predator-like mode can manifest in anything from outright bullying or being pushy, to simply imposing our agenda on an interaction without regard for what the other person cares about.

Such moments can be so subtle they often go by unnoticed. I once was taking a sunset-viewing walk on a beach with a group that included a friend who had just seen the last of her many children leave for college. For the first time in her memory she was now on her own, enjoying the freedom from constant preoccupation with everyone else's needs. She could, say, read a whole book in a day, if she wanted, or travel to places she had never had the freedom to visit.

She was telling me this just as the sun was about to set over the ocean. The light was glowing with the golden hues of dusk when I had the thought that this was the perfect light for a photo. So, inspired to capture the moment, I said with exuberance, "Let's take a picture!" as I clicked open my digital camera.

Immediately afterward I realized that just then she had been gazing at the sunset with a rapt, reflective look. Even so, she went along with my suggestion and posed dutifully as our group lined up for a photo op.

Now we have that moment frozen in time. But when I see it I can't help but perceive it in terms of the predator–prey dynamic. I replay my friend's fleeting look of rapture interrupted, the kind of moment that gets lost in the tumult of our social encounters. And I wonder if my enthusiasm to take that photo—after all, the light was perfect—was subtly predator-like, if only in not first attuning to her in order to sense what worked for her too. Maybe she would have just preferred to continue appreciating the silent beauty of the setting sun.

In horse whispering, the predatory principle refers not to its meaning in the jungle, but to the distinction between using force and sheer control to train and establishing a collaborative connection. That's what Bob meant by putting purpose before connection: we focus on our goals at the expense of attunement.

It's not that spontaneity, enthusiasm, or playfulness should be discouraged. My impulse to capture a moment with my friend wasn't wrong, but I wasn't tuned in to her. We might not realize the subtle emotional messages we send. Taking empathic pauses lets us check whether our enthusiasms or fixed goals are about to override what's happening with another person. Are our motivations shared, or are we making an assumption and forcing someone else's choice?

Any of us can have slight predator-like tendencies, if only by failing to consider how our actions impact other people. For instance, we may be caught up in a hectic rush that's part of our routine and we don't notice how dismissive we might be of others, either of our children or of the person next to us at work. Such subtle controlling interactions go on all the time. Once we start to spot them, it's disconcerting to see how common they are in our lives.

Beyond Melting Down

My husband and I were preparing for a workshop we were giving together. He came into my office and enthusiastically announced, "Here's what we should do," with an agenda of his own.

I quietly responded, "I'm in the middle of thinking through my own ideas. I'm not ready."

At that moment I experienced him as micromanaging. This could have led to triggering my rebellious mode. But I decided to sit quietly and reflect, choosing to take a time-out.

After a bit of reflection I remembered how this was a repetitive mode dance between us: his wanting to take control and my being overly compliant or rebelling. Because this wasn't our first time, I was familiar with some background aspects of this dynamic.

I also know that when my husband and I get excited about ideas, we can both be a little like kids in our enthusiasm. But there's a difference between one kid telling another "You've got to play the game my way" and both kids making up the rules together as they go along. You can see the influence of predator-like thinking in the words of a youngster to his dad, who was trying to break up a fight between the youngster and his brother: "It all started when he hit me back!"

I saw I had a choice: to react to our usual mode triggers, or remember we had a choice not to play out the same reactive mode dance.

So when we met to plan the workshop I started by saying, "It works better when we make these decisions collaboratively," and told him some of these reflections, adding some significant things I was learning that might give us both some insight into the spiral of mode-triggering between us. With a spirit of investigation (rather than indignation), I described all of this to him.

He seemed fascinated and shared his own thoughts on connections he was making. We were both intrigued by this, unraveling an insight into an ongoing, perplexing mystery between us, and we were enthusiastically exploring this puzzle together rather than getting caught up in mode-driven reactions.

It felt like one of those Indiana Jones movies where the guy and the girl courageously enter some "forbidden" zone, like a dark cavern, in the spirit of making some discovery, and then they miraculously discover some long-forgotten jewels and emerge from the danger alive and glowing. Instead of melting down, we were joining up!

This is one of the subtle patterns that often go by unnoticed. But if we can notice the disconnects, which ordinarily occur without our realizing it, we can change course and greatly improve the quality of our communications.

An Evolutionary Upgrade

While I'm working with Bob and the horses, we sometimes get into somewhat philosophical discussions. Bob once told me, "I want this horse to feel safe while I'm working with her so I try to stay attuned to how she might be feeling. If I know she's feeling anxious about something, I want to do everything I can to reassure her so she feels completely supported by this learning. I'm not forcing a horse to do something in a predatory way—we're learning together. That's the art of this work, responding to the needs and changes of the moment and staying tuned in to that.

"The behavioral part of the training is the science," Bob continued. "People focus too much on the science or the technique of learning something without responding to the needs of the moment, which is the art. We should never sacrifice our principles to get another to do what we want—that's predatory."

Letting the means justify the ends, rather than following more humane principles, has led to a long litany of cruelties throughout human history, by people who will do anything to exert control, regardless of how it might harm the connection.

The human version of predator–prey is an I–It relationship, where the other person is regarded as though they were an object, not a person.[3]

This sort of objectification of the other is one of the roots of cruelty; it goes on in training torturers as well as in the spread of hatred between groups. The first step always involves re-perceiving the other as an "It," not as a person.

In contrast, in an I–You relationship, we fully attune to the other person and respond to how what we do or say makes that person feel. The I–You builds on empathy.

"Empathy is the opposite of predatory behavior," Bob said. "If you stay aligned with your principles, with natural laws or a code of ethics, you don't sacrifice your intention to be empathic."

"I probably unintentionally engage in some predator-like actions when I try to get my horses to do something I want," I pointed out. "But I don't feel they mistrust me because of it."

"That's because they know that your intentions are kind and you want what's best for them," Bob said. "But horses are often perplexed by how humans act. They're trying to read our intentions all the time."

"Isn't the point that though the human species can be partly predatory by nature, we can evolve into a more compassionate mode of being?"

"That's the hope."

Swimmy, as one children's story goes, was a tiny little fish among a school of other little fish just like him. They lived in terror of bigger fish, which would chase them and try to eat them. Then one day Swimmy had a brilliant idea: he organized all his friends in the school to swim in a formation that, seen from a little distance, looked like one giant fish. Physical safety is vital too—sometimes alone we may not have that safety, but we can find it in a unified group or community.

When I was six I lived in an apartment in Manhattan near Central Park, which became my playground. One day when my mother, brother, and I were riding the elevator back to our apartment after a day in the park, we shared the elevator with a neighbor, his dog . . . and an adorable baby squirrel. The dog was a ferocious German shepherd with a bad rep for growling with his canines bared and sometimes nipping people. All the neighbors kept their distance. But here we were in the elevator with that mean dog and a tiny squirrel.

The dog's owner explained he had been out walking his shepherd in the park when the dog had suddenly run over to a tree, then carefully sniffed and inspected something underneath. Against his predatory canine nature and in the gentlest way, he then carefully picked up a lone, infant squirrel, and carried it softly in his mouth back to our neighbor.

Then our neighbor asked my mother if we wanted to take care of the baby squirrel. We decided on the spot that we would.

My brother and I played with him for hours every day, giving him my brother's small wooden playhouse to live in. He quickly became part of the family and seemed to forget that he was a wild squirrel. He would entertain us, then curl up in one of our laps when it was time for his nap. I named him Perry.

One day Perry found some walnuts in the kitchen and his foraging instinct kicked in. He carried a few nuts over to where I was sitting in a chair with my hand resting on the arm. Perry lifted up my fingers and tucked his walnuts one by one into the hiding place under my hand.

I could tell countless heartwarming stories about our life with Perry, as his human herd. But reflecting on it now, I'm struck by how that "ferocious" German shepherd seemed to surrender to a tenderhearted compassion and protect this vulnerable baby squirrel. And I was touched by how Perry, a vulnerable prey animal, seemed to feel such safety with us. I find it a hopeful case study in how modes can change: even predators can be kind, and prey can learn to trust.

We are born altruistic, with an innate instinct for justice, the Dalai Lama says, but it can be conditioned out of us. Might the opposite be true too? Can our tendencies toward the predator-like mode be overcome by other kinds of learning?

We upgrade our computers and cell phones all the time. But how often do we think about upgrading our humanity? Could we be in need of an evolutionary upgrade?

CHAPTER

Traps, Triggers, and Core Beliefs

On a beach one day, as I gazed toward the morning horizon and the hues of cobalt and turquoise waves rolling in from the sea, I fell under the spell of a busy yellow jacket in the sand nearby. This feverish insect expends huge amounts of energy and time burrowing its way through furrows of sand to open up deep pathways that seem to have no other purpose than to attest to the yellow jacket's tracks.

I can't help but wonder: what is it searching for? Of course I have no way of knowing what this seemingly pointless activity means in this bug's life nor what compels it to dig away with such intense fervor. Perhaps there is some obscure survival value, all too obvious from within the ecological niche of yellow jackets, to burrowing through sand.

Yellow jackets are a type of wasp, and I've heard that the females of some varieties of wasps dig burrows in sand to lay eggs, where their young can grow to maturity in a safe cavern. But the sand burrows built by these yellow jackets just cave in and get washed away by incoming ocean waves.

I feel concern for these wasps as my mind goes from perplexity at this mystery to what this bug's seemingly poignant adventure says about the ways we humans get caught up in modes of being. We, too, often squander huge amounts of time and energy in wheel-spinning modes that are as futile as that yellow jacket's journey through the sand.

What fuels our misguided drives? What are we looking for? These questions have a most piquant application to our emotional lives—most especially to the ways we repeatedly lose our way.

When we start to see more clearly these futile, relentless patterns at work, we can feel a mix of disenchantment, a yearning for more meaningful pursuits, and a sense of compassion for the poignancy of the human condition. That can be a motivator to wake up, in the ways we've been asleep.

Duty-Bound

Clara was regretting that even though Isabella—a best friend who lived far away—had come to town, they hadn't been able to get together. Clara knew why: she felt duty-bound yet again, unable to seize a chance to have fun.

Here's the story: Isabella was in town on business and had a tight schedule. She called Clara to see if they might be able to have lunch at the hotel where Isabella's meetings were going on, since Isabella didn't have enough free time to visit Clara, who lived about an hour from the hotel.

Clara had been thrilled to hear from Isabella, and loved the idea of them having lunch. Nevertheless, she found herself saying no.

Clara was a creature of habit, and her life revolved around taking care of her three kids and her husband, Tim, a hyper-busy CEO. Behind that story, though, lies another. Clara was the eldest daughter of a judgmental father and an ailing mother, and Clara had learned early in life to give up her own desires when they conflicted with the needs of her loved ones.

Clara was the quintessential perfectionist, which earlier in life had made her an outstanding student and high-performing executive. But as her family's needs grew, she found herself plunging into taking care of her family with the same perfectionist streak. Feeding this inclination, her husband expected Clara to keep the house spotless and well run. Oddly, Clara found this arrangement comforting—it was just like the home she had grown up in.

But when an opportunity came along for Clara to spread her wings and do something outside the house and out of her routine—like Isabella's lunch invitation—something in her said no. And that inability to have spontaneous fun was the nub of her frustration: it no longer felt okay.

Clara's inner naysayer typifies what seems an extreme variety of the prey-like mode called the "demanding parent" by cognitive therapist Jeffrey Young—my insightful mentor and the developer of schema therapy, a pioneering psychotherapy approach based on modes.[1] (In my practice I've integrated schema therapy with mindfulness.) The very name, "demanding parent," points to its childhood origins in strict rules and extremely high standards of conduct. The resulting mode builds around feeling duty-bound and being driven by a compulsive, perfectionistic striving—even at the cost of personal pleasures or health.

Toxic modes are at the extreme end of the negative spectrum, entrapping us in emotional habits that are self-defeating. For instance, while caught in this duty-bound, perfectionist mode, we can become obsessively self-critical and hyperjudgmental toward ourselves. And we hesitate to express our feelings or otherwise act spontaneously—like Clara's letting herself miss the chance to see her old friend, Isabella.

On the other hand, negative modes can also harbor hidden strengths. For the duty-bound mode this includes a strong drive for excellence: having the tidiest house, say, or being ultra-efficient and wasting not a second of time. If we can keep such perfectionistic urges from dominating our lives—keeping us from pleasures or even driving us to exhaustion—then they can help us excel at whatever calling we choose.

But if, like Clara, we get stuck in a negative mode—activating it when it no longer serves a useful purpose or staying in it far too long— then it becomes toxic.[2]

In a toxic mode we find ourselves in the predicament described by the poet David Whyte. Without realizing what we're doing, we slip into telling and retelling ourselves "the same old broken stories" over and over and over.[3]

Mode Extremes

These maladaptive ways of being, like the relationship modes, originate with coping responses we learned early in life. But they often remain strong habits in adulthood long after they have served their purpose. To the extent we become captured by these dysfunctional modes, what may have once seemed a safe place can eventually seem a prison.

In a healthy state, our modes are integrated: we can move smoothly from one mode strength to another (as we'll see in chapter 7). But when it comes to these negative modes, enduring facets of ourselves stay separate, almost as though we were another person while we are in their midst.

The primal template of approach or avoid can take the form of fight-flight-or-freeze responses. In the mind, these emerge as strategies that take the form of distinct modes.[4]

For instance, we can freeze into "compliant surrender" when we feel overwhelmed. People in this mode are at an extreme of a prey-like, overly cooperative stance (as we saw in chapter 5); they surrender and let another person control things, often in an emotional tradeoff to preserve the relationship.

The fight response can take the form of battling against what's hurting or scaring us by overcompensating, bullying, or trying to project an image of being special, when in fact we feel defective underneath. This narcissistic, controlling way of being fits the predator-like mode (again, as we saw in chapter 5).

Another form the fight response takes is the "angry child," in which people express their hostility in the raw, impulsive, uncensored way a child does—an extreme of the aversive mode (chapter 3).

As it happens, Clara's husband, Tim, was a classic example of one extreme of the predator-like mode, where people make up for unmet emotional needs by taking whatever they want from the world.[5] People in this mode can be grandiose, arrogant, haughty, and condescending. They are narcissists who feel free to devalue and control others.

Tim displayed it at work by giving subordinates harsh reprimands when they failed to meet his performance expectations. And at home he was hypercritical of Clara, which only reinforced her duty-bound mode.

Instead of fighting back, Clara would in effect freeze and resolve to do better.

Those in this narcissistic variation of the predator-like mode blame others yet feel they are blameless. Tim, for example, was oblivious to the emotional pain his tantrums caused his employees (just as he ignored Clara's suffering), and he made no connection to the high rate of turnover at his company. In the business and political realms, such people sometimes gather followers—always "yes-men"—and can occasionally rise to exalted positions of leadership; Tim, remember, was a CEO.

Core Beliefs

When Clara was talking with her therapist about her duty-bound pattern she had an aha moment. She had started to monitor her modes, trying to notice the moment when they began, to see what was triggering them.

One day Clara mentioned to Tim that she was thinking of going with a girlfriend to a nearby spa for a day of massage and pampering. Tim was unreceptive to the idea, accusing her of "slacking" just before the holidays, when there was so much to do (at least in his view). As he said this, Clara saw his lip curl in a fleeting expression of disgust.

As she told her therapist, "I felt a pang of fear and self-loathing when I saw that curling lip. And I recognized that look: it was precisely the way my father used to look at me when he would criticize me." Like her husband, Clara's father had been hypercritical of just about everything she did while she was growing up. He never praised her, but pounced on any imperfection with a flood of contempt and sarcasm.

Clara developed the sense of having some flaw in the core of her being that made her unworthy of being loved. This pattern is one of a dozen or so core maladaptive schemas.[6] These emotional patterns act as core beliefs and determine the triggers for the dysfunctional modes. Clara's sense of defectiveness was the trigger for her duty-bound mode.

A mode's core beliefs determine what we will be most vulnerable to, what will not matter, and what will be our strongest triggers. Triggers, like the contemptuous curled lip, are loaded with symbolic meanings that activate a mode. What triggers us is the meaning our minds assign to an event. Something most people don't particularly notice can send another person into a dire mode. It could be as subtle as a certain word or phrase, a tone of voice, or some set of physical or sensory cues.

Consider what happened to a young man who was sitting with friends in a coffee shop. He had been in a positive mood, feeling expansive and happy. They were joking around, laughing uproariously, and he was feeling "on," riffing exuberantly. But when the waitress came to refill his coffee, she poured only half a cup and walked away—and he suddenly felt deflated and sullen. He fell silent, retreating into a brooding shell.

He later recounted this incident to his psychotherapist, who interpreted that sudden shift as due to feeling that the waitress was uncaring. That feeling of not being cared about echoed the young man's troubled history with his alcoholic, self-absorbed mother; when he was a child he felt that she barely noticed him, let alone cared what he needed.

Someone who deeply believes that he is inept or unlovable might unconsciously expect others to be critical or rejecting. Then even the least sign that might confirm these views can activate the related mode.

Seen through the lens of that young man's emotional deprivation— with its core belief that *no one cares*—the half-empty cup triggered a sudden downshift, from a mode of ease and connection to a toxic range of the anxious mode of constriction, where fear and sadness stoke resentful, self-pitying thoughts.

The specifics of our appraisals depend on our particular personal histories and attitudes; we each have our own set. Take that meaning-laden half-filled cup. For someone else that moment might have occasioned a joke or just a shrug. And someone who was trying to wean himself off caffeine by gradually drinking less coffee—a very different mental model—might even have thought the waitress had done him a favor.

"The difference between a good day and a bad day is your attitude," as a saying goes.[7]

A mode's core beliefs—its underlying attitudes, outlooks, and perceptions—act like mental algorithms, unconscious rules that control our emotional decisions.[8] These work along the lines of "if . . . then." For example, when someone senses that her partner might be distancing himself, in the avoidant mode she might make a sudden, preemptive exit; but if she were in the anxious mode, she might become demanding and clinging. The decision rule: *If* I feel distanced *then* I will be abandoned and must protect myself—and the mode determines which strategy we tend to use.

As another example, an extreme form of the aversive mode (something like what Young calls the "punitive parent") orients people around the belief that anyone—including oneself—should be harshly punished for making mistakes. This mode's moralistic and intolerant point of view makes people overly harsh. Excuses, extenuating conditions, and ordinary imperfections do not matter; the mode lacks empathy for the person being condemned. Forget mercy, let alone forgiveness.

One intriguing giveaway of a person caught in the punitive zone of the aversive mode: that person's tone of voice often mimics his or her own parents' scolding—say, cold and full of contempt. Each mode speaks with a unique voice, more frequently in our inner dialogue than actually.

The inner voice expresses the mode's algorithm regarding our world, a fixed set of assumptions that become perceptual templates for us. In general, the brain acts as a hypothesis engine, updating our beliefs about the world around us. But a mode sticks to the same beliefs without updates.

These personal theories or beliefs guide our sense of what matters and what does not, and what it means—and therefore how we interpret a range of experiences. They are built around self-referencing, a myopic view that leads us to see encounters as "all about me." We tend to see ourselves on center stage, with whatever happens referring to us. As a friend put it, "We create an internal movie where we are the star of our own film, the producers, the directors, the whole cast of characters—even the critics!"[9]

Core beliefs are self-fulfilling prophecies. They impose their assumptions about the world, selectively taking in any information that supports their view, and ignoring or discounting anything that counters these views. While under the spell of a maladaptive mode our perceptions skew so we misinterpret what's actually going on.

As these patterns are enacted again and again over the years, core beliefs can become fixed attitudes: solidified, habitual, and automatic, and harder—but still possible—to surface and change.

Triggers

A well-dressed woman wearing elegant jewelry came to one of my workshops. The jewelry she wore, it turned out, was of her own design—and it had a poignant story.

A few years back she had become a jewelry designer, and within a short time her work had attracted notice in the press and was carried by some high-end boutiques, even in the shop of the Museum of Modern Art. One day she got a call from a buyer of a famous department store who wanted to see her jewelry. The buyer loved her designs but asked for some modifications that would take a little time. They made an appointment to meet again to see the modified designs and pin down the details of what could be a large order.

Four months later the designer went back to the department store, only to learn that the buyer who had loved her work had left the company; a different buyer had taken her place. That new buyer took a cursory look at her jewelry and, with a curt wave of her hand, dismissed it as "uninteresting."

"I was crying as I walked through the narrow hallways of the store's business offices," she said. "I was thinking, *This is the Hall of Shame.* I felt like an impostor who had been unmasked."

In the grip of her anxious mode, in the form of intense self-doubt, she went straight back to her studio, melted down all her new designs, and never made a single piece again.

But now, a few years later, she was learning about the emotional patterns that trigger modes. The jewelry designer recognized as her trigger a core belief: fear of failure. "Despite my success," she said, "during the whole time I was a jewelry designer, I always felt like an impostor. I had no training in jewelry-making, and I was afraid I'd be exposed as a fraud."

Triggers generally are events or cues that launch us into a mode. These moments in life take the reins out of our control and lead our minds down a path of confusion.

For instance, an e-mail from a mildly irate client sent an interior decorator into a flurry of self-doubt, anger, and worry—an anxious mode. "For hours I've been going over and over that e-mail," she told me. "Was I right? Did I do something wrong? All I have to do is think of her name and my mind is spinning all over again. There's a lot of anger. It's all so familiar."

Frequently modes are domain-specific, only active in certain parts of our lives. Someone can be avoidant in relationships and anxious when it comes to health. A boss can be tyrannical at work—hypercritical and demeaning—but an angel at home. Or a particular person, like a certain family member, might trigger a mode in you that does not come up when you are with other people.

The fear-of-failure trigger is found in a list of schemas, each of which embodies specific core beliefs that harbor a unique set of triggers.[10] Exactly which mode a given core belief activates depends on our particular life history and how we've learned to cope with the unleashed feelings.

Here's a list of common core beliefs and the modes they often trigger:

FEAR OF ABANDONMENT (*I'll be terrified if I'm alone*) makes us hypersensitive to the least sign (even imagined ones) that someone we're attached to might abandon us. This belief may trigger anxious clinging or avoidant modes.

EMOTIONAL DEPRIVATION (*No one cares*) is the belief that our needs are being or will be ignored in our adult relationships. We become hypersensitive to any sign that this might happen. This belief may trigger anxious or avoidant modes, or sometimes an overcompensating sense of entitlement.

SUBJUGATION (*My wishes or feelings don't count*) leaves us passive and unable to assert our wants (and uncertain of our needs) in relationships. In a prey-like mode, we wait to be told what to do and surrender to the wishes of others.

SELF-SACRIFICE (*I must put everyone else's needs first*) melds the beliefs of deprivation and subjugation, so we sacrifice in order to put everyone else's needs first. This triggers a prey-like compliance mode.

UNLOVABILITY (*I am deeply flawed*) is the belief that we are defective at our core, that no one who really knew us could love us. It gets triggered by signals of rejection or disapproval and can, in turn, trigger a range of modes, including perfectionism.

SOCIAL EXCLUSION (*I don't fit in*) is the feeling that we do not "belong" in a group or will be ridiculed. It triggers the anxious mode, when we are in a group of strangers or in crowds.

VULNERABILITY (*Bad things will happen*) is the fear that we are always in danger somehow. It leads to the anxious mode. We imagine catastrophe and exaggerate danger at the least sign of threat or risk, avoiding what makes us anxious. This can be about money, health, or the safety of the people we love.

UNRELENTING STANDARDS (*I must do better at all costs*) is the belief that we must strive to meet impossibly high standards and sacrifice oneself in order to achieve anything. It can apply to any domain—sports, school, career, housekeeping, and so on—and can be triggered by a performance measure or a comparison. It activates the duty-bound perfectionist mode.

FEAR OF FAILURE (*I don't have what it takes*) is the conviction that we are not up to some performance challenge. It can activate the anxious mode when we are faced with that demand. The triggers are similar to those for unrelenting standards, but the difference in core beliefs (*I can do better* versus *I can't do it*) leads to activation of a different mode.

ENTITLEMENT (*I'm special*) is the attitude that because we are due special privileges, ordinary rules do not apply to us. Plus, we can do whatever we want, because others' feelings don't matter. It activates the predator-like mode.

If you watch boats moored near each other in a harbor, you'll notice that they all point in the same direction, aligning and adjusting to the winds and currents. As time passes, the boats will shift where they are pointing, again in unison. The forces that cause this change in orientation—the shifting winds and currents—may not be visible to us, but their results are evident. This, too, is how triggers activate our modes.

The Evolution of Emotion

A nursing student was paying her way through school by working as a server in a restaurant. "I've always been motivated to become a nurse," she said, "because I want to take such good care of people that patients say, 'Even though that was a tough medical procedure to go through, I had a nurse who was so kind it made all the difference.'"

She's an unusually thoughtful server, noting when customers need something before they have to ask. "Being a kind waitress is good training!" she says.

In a secure base, one has the sense of being enveloped in a nurturing inner cocoon. The compassionate medical care the waitress envisioned means creating a shared secure mode, where people feel comforted and protected, even as they confront what may be harsh medical realities.

Children form a trusting outlook, or working model of the world, if their experience has been that people are caring, that the key people in their lives will be attuned, and that they themselves are worthy of love.

The core beliefs underlying a secure base lead to feeling positive in one's closest relationships, and in one's life in general, making a person calmer and more confident, and more able to establish intimate, caring connections.

The primal model for this secure mode lies in the dependence we all feel in life's earliest years. We're born into a state of helplessness, and we need protection and nurturance to survive. In a protected, caring environment—a secure base—an infant acquires the deep sense that the interpersonal world is a safe place, and that people tune in to your feelings and needs and help you with them.

"Happy soup" is a pleasant memory from my childhood. My grandmother would say, "Come over to my house, and I'll make you some happy soup."

"What's in happy soup?" I once asked her.

"It doesn't really matter," my grandmother replied. "As long as it's made with love, it's happy soup."

And when I was at my grandmother's having happy soup, that sense of being well loved made me happy too.

As this sense of safety and connection carries over into a secure base of adulthood, British developmental theorist John Bowlby says, it creates a "safe haven" for exploring the world, ideas, and relationships.[1] This sense of security lets us see ourselves as valuable and lovable, with a sense of self-worth that grows from all the times we felt emotional need and found ourselves cared for by an empathic loved one—unconditionally loved.

But there's an important distinction between an inner sense of a secure base and being with people who give us the feeling of a safe haven. With an inner secure base, we turn to our own resources; we don't depend on others to prime this mode.

I use the terms "secure" or "integrated" to refer to a mode with an overlapping set of positive qualities, which are highly adaptive; these include Bowlby's but go beyond. This mode can display a different face according to which aspects get highlighted at specific moments. At times it might be apt to call the combined mode adaptive, positive, or resilient, or label it as basic goodness. It might be better for the mode's name to change with the circumstances, or with different people, at their different levels of development.

The multiple dimensions of this positive range of experiences are understood differently in various schools of thought. Which dimension is highlighted depends on what lens we look through. In addition, we all relate to this positive mode differently, since each of us has perhaps developed some aspects while other aspects still need strengthening.

A sampling of the positives in human nature gives us a fuller picture of this mode. Together they enhance a well-integrated development of the whole person—psychological, physical, spiritual, interpersonal, and altruistic.

For instance, evolutionary psychology emphasizes the ability to adapt. Clinical psychology highlights the importance of emotional adaptability. Healing arts, like qigong, talk about the unobstructed flow of energy and balance.

Positive psychology focuses on happiness and having a healthy attitude, while work on flow and optimal performance describes a mode where we perform at our best and feel joy. Cognitive science looks at this mode in terms of heightened mental efficiency.

Systems theory emphasizes the interdependence of people in relationships and with the planet, which widens the frame around this mode. In the social dimension the focus is on attuned connections and a sense of "we," while those concerned with social and environmental issues put it in terms such as having an attitude of selfless service, doing compassionate work for justice and reconciliation, or finding sustainable solutions for the planet.

Young's schema model has an extensive description of this mode's aspects (which he calls the "healthy child" and "healthy adult"): we feel "loved, contented, satisfied, fulfilled, protected, accepted, praised, worthwhile, nurtured, guided, understood, validated, self-confident, competent, self-reliant, safe, resilient, strong, in control, adaptable, included, optimistic, spontaneous."[2]

While a Western lens highlights the qualities that surface with integrated wholeness or enhances the positive states, an Eastern lens might focus on removing what obscures our genuine nature. Buddhist psychology tells us that everyone has at his or her core a "basic goodness"—something like an inner secure base. Qualities of a happy and free mind—such as confidence, love, equanimity or a sense of ease—are part of our core nature; we can turn toward these natural inner qualities and can trust in our abilities to nurture our basic goodness.

The secure mode, like mindfulness, brings the inner world into balance. These healthy ways of being dissolve the less healthy ones, like sun melting away the clouds. The secure mode does not require that we have had a perfect, stable childhood. We connect with these healthy states during our more open-hearted, clear-minded moments. Still, there's no magic button that I know of, no surefire way to create some everlasting comfort zone.

Robin's Story

Robin's family has a large spread with a herd of horses. She practically grew up on her horse, which she often rode bareback, and has always loved animals. She wanted to be a veterinarian, and while interning with a vet on a Navaho reservation in Arizona, her bareback riding talent got her a spot in a local race.

She spent much of the next year prior to the race practicing her bareback riding. Her internship at the vet had ended, but the following summer she returned to Arizona to dedicate four weeks to racing for her friends there, a Navaho family. She had been there just a few days, preparing for the race. One of her hosts had taken her on a fantastic trail ride through the stunning Arizona countryside, and that night, as she and three members of the family were returning home in their pickup, they were hit head-on by a speeding van. Robin suffered massive injuries; two others in the car were killed. "The summer of my dreams," as Robin says, "turned into the summer of hell."

The impact broke both of Robin's legs and an elbow badly, and lacerated her liver. Adding to the trauma of her friends' deaths, police said the other driver had been on a suicide mission in a stolen van.

One moment, Robin recalls, she was laughing with these new friends, and then "suddenly they were taken away. It was totally weird, hard for me to wrap my head around. Except that I take the positive spin and say, 'This was not a personal attack; it was a random act.'

"While you're going through something like this, in the depth of the situation, you're in survival mode, just getting through it," Robin reports. "Luckily for me, survival mode means thinking about the positive. We have to look at it in a way that is healthy and makes everything move forward. Because when something like that happens, and you're physically and mentally injured, you don't need anything else to move you back. It really helped to revision myself as being normal and healed—seeing myself as getting better."

Robin's resilient attitude in the face of adversity signals a secure mode, where her core beliefs included a deep conviction and inner resources that she was able to draw on. Seeing the world through this positive lens allows hope and strengthens resolve.

In secure mode, we have a sense of stability while we adapt to changing conditions. Scientists say that the thickness pattern of tree branches makes them maximally flexible in wind, so they are far less likely to break. In this mode, we're able to adjust with resilience and flexibility, like a tree bending in a high wind.

This is not to deny the harsh struggles people go through, nor to make it all seem too easy. Some realities are extremely hard to bear. Some pain needs to be expressed and calls out for loving support and care. We each have our unique timing and ways of dealing with life's hardships; grieving and loss have their own natural rhythms. The path through suffering is not always so clear. But we each need to find our own way in the face of our challenges and the difficulties we face, and to care for ourselves and each other in whatever ways are needed.

I asked Robin if her accident and recovery had changed her outlook. "Now I understand that once you've been hurt and you heal—or your heart is broken and you get over it—it's empowering. It's a feat that you were able to recover. Getting into an accident like that is so random and strange. It was a reminder that life can be extremely volatile, and bad things can happen to anyone.

"I just had to accept that I could work through this and not take it personally," Robin recalled. "Not be sad and feel sorry for myself. Not 'Why did this happen to me?' That would be poisonous. I would dwell on it and I would be very unhappy. So I decided that I would just stay true to myself. I don't regret things. I don't regret going out to Arizona and having this experience. It would be damaging if I thought I shouldn't have gone out there, that I did something stupid by going toward a dream."

Our sense of security lets us feel safer, more loved, and well supported amidst life's turmoil. Robin displayed many of the courageous attitudes typical of the secure mode: gratitude toward those who have helped you in life, even in small ways, and hope when things are difficult.

A base in positivity can ease our suffering—it's not that we can keep suffering at bay, but we might keep it from being a negative force when it comes.[3] "There's a crack in everything," sings Leonard Cohen. "That's how the light gets in."

Authentic Connection

A year after her dreadful accident, Robin told me, "One way this has changed my life is really appreciating other people and letting them know that. I love my friends. I love my family. But now I focus on *saying it*. Saying 'Thanks for your help today,' or some such that I may not have before. Make sure they know the warmth that you feel for them. You have nothing to lose. It's something people should do more often."

That kind of authentic communication flows more naturally when we can speak from the heart in ways that reach deep into another person, resonating with that person's genuine nature. And we do so naturally from our own inner sense of a secure foundation.

Joining up in authenticity is a two-way street. Once she had returned home, Robin found that many people were reluctant to talk about her traumatic accident. They would skirt the topic even when they saw she was in a wheelchair.

But not the kids. "Kids don't have that filter where they think, *I probably shouldn't ask her about that*. Kids would just ask me questions, and at first I wasn't prepared for it. One little girl I used to babysit for said, 'So my mom told me you almost died. Can I see your scars?' My reaction was, *Okay, this is getting a little bit touchy*. But I wasn't mad about it. She was just being curious. She was asking me the questions that everyone was wondering about but not asking me directly. She was able to say what the adults wouldn't. It was actually quite helpful, because I was able to see that she didn't think this was a big deal."

Robin added, "So I decided, I'm not going to get closed up about this—but not tell her anything she'd find scary. I'm going to tell her the truth in a way that is appropriate, that makes it positive, that everything's okay.

"When I was going through my life as a student and working and rushing around, I had my blinders on, and not a lot of people were telling me that they appreciated me. Then when I was in this compromised state, I saw that people could be very, very helpful. So often it's hard to stop and take the moment needed to put yourself in someone else's shoes. It came down to these people giving me extra oomph, which helped me appreciate what I could give. That's what gave me the nudge to start my emergency medicine training. I want to be [a] master EMT so I can be a first responder."

Having a secure inner base lets us join up naturally. A genuine connection lets us be sensitive to other people's needs as well as our own, fostering empathy, caring, and concern—the basis of compassionate action.

Positivity

Ted's father had died after a long illness. The death was not a surprise. As Ted put it, "I said everything to him I needed to. We were at peace."

Still, Ted was surprised at how much turmoil he felt afterward. "I just didn't expect to feel this upset." That turmoil was intensified by the turbulent interactions he had with his siblings. His father had left a will regarding his financial assets, but he had neglected to say how his personal possessions should be divided. Some were quite valuable, and Ted and his siblings found themselves wrangling over who should have what.

The hardest thing for Ted was the pressure he now felt from his family to step into his father's role as president of the family business. Ted was the only one of his generation to have any business experience; he enjoyed running his own business.

"I'm at a complete loss about what I should do," he told me. "I might want to take over my dad's business, but I don't want to make that decision just because people say I should. I can't find my North Star by listening to shoulds. I need to tune in to my heart, listen to my gut. And

I'm so upset these days, I just don't feel I'm in the right frame of mind to make such a decision."

Ted, a devout Catholic (who in his younger days had thought of going to seminary), had a contemplative prayer practice he relied on to find his true center; it was this wiser mode he longed for now. "But I just can't calm my mind these days. There's too much that upsets me."

After we talked it over, Ted decided to take time off over a three-day weekend and go to a Benedictine retreat center he had loved during his student days. As he put it, "I just need to jump-start my calm center. Once I have that back, I can make up my mind from a better place."

He was reaching for a positive mode, where we make our best decisions. While in the grip of a distressing mode, the amygdala captures the brain's executive centers in the prefrontal area and the mind constricts, seizing on overlearned habits and failing to entertain more flexible, imaginative options.

In contrast, from our secure mode we can entertain alternative solutions to life's problems. The mind operates creatively; the world of possibility expands. It's like in sailing: you can control the boat by adjusting its sails—that's our basic security—but you need to yield to the winds—be adaptable.

When a person is in a positive frame of mind, scientists see increased activity on the left side of the prefrontal area. But when distressing feelings run too high, the brain is wired so that the prefrontal area gives up control to the emotional centers, particularly the amygdala. During intense negative emotions, brain scans reveal heightened activity not just in the right amygdala and its connected circuitry but also in the right side of the prefrontal cortex. The amygdala's emotional urgency hijacks the prefrontal area and takes over.

Maladaptive modes likely run in the same bottom-up fashion, with our emotional centers seizing some control. In contrast, while we are in a positive mode, the prefrontal area remains firmly in charge. We have maximal flexibility in our attention, our thoughts, and our insights—our fundamental abilities to respond to the world effectively.

In a positive mode, our feelings can go beyond a wider ranger than happiness and joy. They can include a serene composure, a keen interest that lets you get fully absorbed in whatever is at hand, a readiness to be amused at the unexpected, a sense of awe at a stunning sunset, a heart-warming sense of appreciation and inspiration at seeing an act of kindness or a musician's remarkable performance. And, of course, love in its many-splendored varieties.

Healthy, adaptive modes are the best platform from which to embark on the tasks of a full and engaged life, such as being responsible and committed, going to work, parenting, taking care of our health and staying fit, pursuing pleasures like connecting with loved ones and aesthetic delights, and diving into intellectual and creative explorations.

Such fruits of positivity have come under the microscope of science in research led by Barbara Fredrickson, a psychologist at the University of North Carolina. She finds that if we keep track of all the negative and positive feelings we have during a given day, we can get a ratio of positive to negative that gives us a good idea of our overall capacity for positivity in our lives.

The average ratio is two positive feelings for every negative one. People who are depressed have the reverse ratio: two negative feelings for every positive one. But with an enhanced positive attitude—a ratio of three to one—we hit a tipping point where we break out of the ordinary ups and downs of life and start to flourish.

At this point we go through a phase transition, a shift into the range of positivity. In this mode range we more readily enter the "flow" state, where we perform at our peak, handle most challenges well, lose ourselves in what we're doing, and feel a sense of wellbeing and an easy openness to connections.

Positivity does not mean passive acceptance, the blind assumption that everything is fine just as it is. As the Dalai Lama notes, "Realistic fear is necessary. Wisdom means awareness of all aspects of reality. But with strong disturbing emotions you can't see all this."

Of course hardship can sometimes lead to stress and even depression. To prime the secure mode in such times, we need what the Dalai Lama calls "helpful inner circumstances," capacities we can cultivate, such as adaptability, calmness, equanimity, lovingkindness, or insight, to mention a few.

Fredrickson's research finds that positive emotions, such as joy, serenity, and gratitude, broaden awareness; we literally can see more. Positive feelings are the default mood range for this mode, but not rigidly so. An adaptive mode lets us respond freely with whatever emotions are apt for the moment, whether grief and sadness at a loss or joy on hearing good news. We are genuinely ourselves. We are also more able to express constructive indignation in response to an injustice or a thwarted goal. And, as Aristotle said, while it's all too easy to get angry, we usually find it hard to get angry at the right time, with the right person, in the right way, and for the right reason.

Our brains seem hardwired to send and receive positivity; we have a class of neurons dedicated to the sole task of spotting someone's smile or laugh, instantly making us smile or laugh in return. This makes positivity contagious and playfulness a powerful mode shifter.

As the 17th Karmapa, the Tibetan teacher, says, it is important to be playful while awakening.[4] Having fun is a sure sign that people are in a shared secure mode, where we inhabit a paradigm of peace. Humor and the arts create a social bond that has been a connective force throughout our evolution.

Once I was at a board-of-directors meeting of the Seva Foundation. It was held outdoors on a farm in Northern California, and we were seated in a circle amidst the large teepees that the following week would serve as dorm rooms for the kids going to Camp Winnarainbow. The day was dry and hot, and torpor was setting in. Some people were perched on the ground with their eyes drooping behind their sunglasses or their heads leaning on their hands; others relaxed, lying on their sides. People were finding clever ways to look attentive, but the "board" meeting was slipping into the "bored" meeting zone.

Some very large birds were lazily circling high in the sky above our group. Seizing the moment, founding board member (and clown), Wavy Gravy, shouted an urgent warning: "Turkey vultures! Look alive!"

That woke us all up!

Finding Our Strengths

A friend I often rely on for editorial feedback can be brilliant in her insights. But she is also prone to the perfectionist mode. This can lead her to make overly judgmental comments, which I don't find helpful.

She had been reading drafts of my manuscript on mode work and started to try this work herself. She had become familiar with the modes, so one time when I gave her a section for her comments, I jokingly said, "Please edit from your secure mode, and add a bit of critical discernment from your perfectionist mode. Leave out the judgmental part."

While she clearly saw her own perfectionism, she was taken aback to realize that she was bringing it to her editing. "Once you're aware of your own patterns," she said with a laugh, "you can be surprised by them. I had no idea I did that."

Modes are mixes of habit, with positive and negative poles. In the perfectionist mode, for instance, there can be a sharp, intelligent discernment at one end and a harsh critic's judgmental opinions at the other. If we set aside the parts of modes that are counterproductive, the adaptive parts represent strengths.

These strengths are the parts of our habit repertoire that can represent our positive capabilities, if they can be freed from the distorted perceptions and overreactivity of the rest of the mode. This can happen as we access the positive mode range.

For the avoidant mode, the upside might be a knack for inhibiting upset and being self-contained. For the anxious mode, the emerging strengths might be a knack for ensuring a sense of safety or for staying connected.

In the anxious mode we ruminate about things, such as needing to know if we are safe or if our connections are stable, or just ruminating to gain a clearer understanding of some issue. But if we can use this same inquiring, investigative quality of mind—a more spacious reflection without the overlay of anxiety—it becomes illuminating rather than ruminating.

Likewise, the core beliefs that trigger the maladaptive modes have potential strengths: Subjugation can transform to assertiveness. Deprivation can turn into a nurturing, empathic sensitivity. Feeling unlovable harbors a genuine humility. Within entitlement lurks a healthy confidence. Perfectionism becomes competence and efficiency. Abandonment holds loyalty. And vulnerability holds a healthy cautiousness.

There are also strengths in each of the "three roots," the Buddhist modes. The better pole of the attachment mode, for instance, can give us a well-honed aesthetic sense, inspiration, joy, compassionate motivation, emotional attunement, and trust—all highlighting the positive qualities of people. Attachment can transform into an energetic drive and a discriminating wisdom that helps us accomplish great creative works or compassionate pursuits, and can be harnessed for awakening—the desire to be free.

The aversive mode's beneficial aspects include what Tibetan Buddhism calls "mirror-like wisdom": a keen clarity, a sharp discriminating intelligence—as with my friend's edits—and healthy discernment. There is a potential for working creatively with conflicts, for example.

And the bewildered mode can yield equanimity, wise spaciousness, and disillusionment with our pointless habits.

The predator-like mode's positives include a strong confidence, while the prey-like mode has ample empathy and a talent for collaboration.

For each of these modes, the negative poles manifest when the mode mixes with emotional agitation and distorted perception; the positive aspects gain strength as we transform them. Rather than assuming everything in a negative mode is something to be rid of, we can acknowledge the mode's strengths and enhance them.

Full maturity allows us to integrate capacities into our being that have so far eluded us. For someone prone to the avoidant mode, for example, that could mean becoming emotionally attuned and expressive. And for those given to the anxious mode, it would be finding equanimity despite disturbing feelings.

To the extent these parts of ourselves have been split off into the repertoire of a negative mode, healing means giving us full access to their positive capacities and integrating them into our being. Another way to think of this positive range of being is as our integrated mode, where all the fragmented and disconnected aspects of ourselves, manifested in our negative modes, offer up their strengths and become accessible to the rest of our being.

Jake was the school bully, the predatory eighth-grader everyone else feared.[5] One spring day, two classes—Jake's and a much younger class of third-graders—had been on an outing together. On a remote country road, as their buses were heading back to the school, they ran into a powerful storm. The winds were dangerously high. The bus drivers were looking for a safe place to park when the bus with the third-graders turned over. Thankfully no one was seriously injured, but they all had bruises and bumps and all were crying.

The older kids jumped out of their bus and ran to help. In the chaos, one teenager emerged as the leader: Jake. He directed the other eighth-graders to each match up with a younger kid to help them keep dry, calm them down, and get them all to the safety of the other bus.

Jake had drawn on the strengths of his usual predator-like mode to become a confident leader, using those strengths not for his selfish ends but in the service of compassion.

Recognizing Modes

Let's say you think you've lost something because it isn't in its usual place. It has great sentimental value to you, and you automatically assume, with a sense of panic, that *it's lost!* Of course it's natural to feel concerned when you've lost what you treasure, but fixating on the assumption and envisioning catastrophe—perhaps without even looking to see if you might have put it somewhere else—strays into the distorted mode range.

But if you take a moment to think twice, check your assumption, and calm down (*maybe it's not lost; maybe I should see if I put it somewhere else*), then you don't enter the negative mode range.

These two moments are microcosms of two very different modes. In the first, your thoughts might have begun to follow along the rut of a negative mode habit, particularly in making the instant assumption that because you couldn't find it the first place you looked, it must be lost. Such alarmism, along with that rush of panic, is a sign of the anxious mode. The second moment reflects the secure mode, where you're able to think more clearly and question the distorted assumption that the item was lost.

When you know the flavor of a mode—the habitual thoughts, feelings, actions, and interactions associated with it—it's easier to spot a mode the moment it activates.

In the *anxious mode*, we make distorted, negative assumptions that exaggerate our worries and too-hastily confirm our fears. Our focus fixates on the worry. The feelings that come along with such thoughts are anxiety, fretting, and a nervous energy manifested in the body. The impulses to act might include acting out of fear, perhaps by being overly cautious. And our interactions will tend to focus on our anxieties.

By contrast, in the *secure mode*, we're more understanding, have a larger perspective, and can think more clearly. We question faulty assumptions. Problems seem more workable. If we get upset, we recover more quickly. We feel stable, centered in ourselves, and replete. We're more responsive and less reactive. We feel connected, even joined up, in our interactions, and are more kind and patient. We view things more positively.

To recognize what mode you might be in at a given moment, the first questions to ask yourself might be general ones: Am I in a negative or positive mode right now? Do I recognize any mode signals in my own mind or in someone else's?

General signs of the negative mode range include:

- Distorted thinking, ruminations, or confusion.
- A volatile emotional reactivity tinged with negative emotions, such as fear, anger, or shame, or a sense of numbness instead of what could be considered appropriate feelings.
- A shrinking attention span. The mode reactions fill the space of awareness, crowding out everything else. We perceive through the mode lens.
- Other activities being shunted aside as the mode habits dictate what we do. We have impulses to do something we may regret later. We're defensive.
- Our interactions with others narrowing to the issues of the mode.

Once you clarify that you are in a negative mode of some sort, you can go a step further and identify the most likely specific mode you are in. You can use the descriptions in each chapter or use these thumbnail sketches for quick guidelines.

The signals of the anxious mode and the secure mode were just detailed. Here are brief summaries of some main characteristics of the other modes we've described:

ATTACHED (*I want*): Greedy yearning. Grasping and clinging.

AVERSIVE (*I don't want*): Negativity, anger, and resentment. Sees faults, makes harsh judgments, and rejects.

BEWILDERED (*I don't understand*): Feelings of confusion or agitation. Indecision or indifference.

AVOIDANT (*Stay away*): Numbness. A withdrawal from people and an avoidance of emotions.

PREDATOR-LIKE (*I'm the boss: I'm special*): Extreme confidence or grandiosity. Arrogance, condescension, or a lack of empathy.

PREY-LIKE (*I must do whatever others want*): Helplessness. Passivity and an acquiescence to demands.

PERFECTIONIST (*I must meet the highest standards and do my duty*): Feelings of guilt and self-judgment. A duty-bound lack of spontaneity.

SECURE (*I am safe, capable, resilient, positive, connected*): Feelings of security, repletion, happiness, and openness to others. Acts with confidence and flexibility.

Recognizing our modes when they activate—noticing how they make us think, feel, act, and interact—is half the work. Bringing these otherwise invisible habits into the light of awareness allows us to begin to reduce their hold over us.

Mind Whispering

CHAPTER

8

Shifting the Lens

I remember once being in New Delhi, where the enormous volume of traffic can create an unimaginably tangled mass of trucks, cars, scooters, motor rickshaws, and the occasional cow. At the crossings of major boulevards, the traffic lights can take many minutes to change because of the endless sea of vehicles that have to pass through.

My husband and I were in a taxi with the Tibetan lama Tsoknyi Rinpoche, on our way to get some supplies for a Mind and Life meeting we were all going to attend in the hill town Dharamsala. As we were inching slowly across the city, we were waiting for another seemingly endless red light to turn green. Worried about all the delays making us late, we were getting a bit annoyed at how long the light was staying red.

Then we noticed a word lit in silver and embedded in the red light, which urged: RELAX! When we saw the message, we laughed so hard it actually *did* help us relax.

Later that evening we shared another taxi on the way to the Old Delhi station, where we would catch a train for that meeting. Traffic was still a tangled mess. The jammed streets were keeping us at a standstill with only an occasional break letting us bump ahead a few feet. In Indian traffic, cars can practically be on top of each other; yet none of the drivers seems to think twice about the hopeless patterns of gridlock they help create.

Worried about missing our train—and the first day of the meeting— my husband and I were getting tense, even more so than earlier in the day. As the traffic lanes got more jammed we could feel ourselves heading into an anxious mode. Just as panic was taking hold, Tsoknyi Rinpoche said good-naturedly, "Just watch. Everything will untangle and we'll get to our train on time. Relax!"

That timely reminder was once again the perfect antidote—this time to the anxious mode. And to our surprise, the jammed lanes miraculously unraveled like a ball of yarn. We made it to the train with time to spare.

The point of mind whispering is not to attain some blissful or special state while sitting on a cushion, but to be able to deal calmly and clearly with traffic jams and bumps on crowded streets, with difficult people and intense relationships, and with the boisterous chaos of our own minds.

When we are in the grip of a negative mode, whatever problems arise loom large while our view of options narrows. We tend to focus on a single cause rather than the web of underlying conditions. We may rush to blame someone.

You can't always change external realities, but you're always free to change your perspective. Sometimes a gentle mental nod—like that RELAX!—is all it takes to prime a change from panic to laughter, and so shift to a secure mode.

Mindful Habit Change

When I was a teenager, my mother and I often had engaging psychological discussions, and I appreciated her openness and insightful perspectives. It sometimes helped to talk in theoretical terms, depersonalizing our own interpersonal "madness." Once, several years ago, we locked horns about some issue. She complained defiantly, "You can't change people!" And I responded willfully, "You can't change people, but you can change patterns!"

Suddenly, as if a light had been switched on in both our brains, we continued examining things through a psychological lens and forgot that we were fighting.

It's like the lightbulb joke: How many psychologists does it take to change a lightbulb? Just one—but the lightbulb has to want to change. It's the same with modes.

Modes, like all habits, break down into three basic parts: triggers, routines, and rewards. Triggers are the cues that initiate a mode. Routines are all the habitual patterns of thoughts, feelings, and behaviors we engage in while in a mode. Rewards we feel from within; for example, a negative mode might be simple relief from an even stronger negative emotion.

The basal ganglia system—the part of the brain that manages our habits—follows simple decision rules. First, it looks for a trigger: the cue that tells it what routine to apply. When you see yourself in the mirror first thing in the morning, your basal ganglia network tells you to wash your face and then reach for your toothbrush and toothpaste. Then it registers the reward that such a routine elicits: the pleasant feeling of being awake and refreshed. This strengthens the brain's circuitry for that habit.

The triggers for modes range widely but often have a powerful symbolic meaning in the mode's perspective. Something about a tone of voice or certain words, say, can resonate with the original cue in ways that make them a mode trigger. Such cues elicit the learned routines that make up the mode.

For the avoidant mode, for instance, overwhelming emotions are triggers, and the mode response includes all the ways the person withdraws. The reward that reinforces such a mode can range from a physical sensation to an emotional payoff. For the avoidant mode, the reinforcement comes from lessening anxiety and agitation. While withdrawing from strong emotion and from people has obvious costs, the emotional relief lights up reward centers in the brain, which cements the mode habits into place. We cling to these automatic habits, whether they make sense or not.

Here's what such a shift looks like. My friend KD knew the origins and signs of his punitive pattern: "I left home when I was eighteen, but I brought my parents—and their harsh parenting—with me inside. I was horrified to see myself repeat my parents' behavior, to hear their tone of voice coming out of my mouth. I lived my life protecting myself against the darkness and unhappiness that I expected to come at any moment, and was ready to pounce.

"I was once really upset about something and when I came into the house, my teenage daughter was sitting in the kitchen. She hadn't cleaned up the dishes. As I stormed through the kitchen, I snapped at her the way my mother would have snapped at me."

The cue—the messy kitchen—triggered his habitual punitive reaction.

But then came a change for the better: "I'll never forget the way she looked at me. She was astounded that I was in such a weird space. She didn't take it personally at all. She just looked at me like I was from some other planet. And because she didn't take it in, by the time I had finished walking through the kitchen, I saw how stupid I was acting and dropped the whole thing."

Waking up to his toxic pattern was rewarding in itself. The key to changing modes lies in examining the cues and routines that drive them and replacing self-defeating reactions with more healthy ones, which have their own more positive reward.

Mind whispering integrates mindfulness with habit change. This integrative mindfulness provides the awareness that a mode habit needs to change in the first place and the ongoing reminder to practice the new, positive alternative.

Bringing our automatic mental ruts into awareness is the first, crucial step. If they stay unconscious, we're powerless to do much about them. Only by recognizing a negative mode as such can we begin the process of habit change.

In the next step, we intentionally replace the habitual responses with more productive ones. And we practice that shift at every naturally occurring opportunity—whenever we realize the mode has begun or recognize one of its cues.

Our daily lives are peppered with cues for our modes. Going on vacation or on retreat can temporarily clear the slate by freeing us from the influence of our habitual cues and letting us begin new routines. Mindful awareness cultivates this open, on-vacation quality that lets us more freely start new responses. This fresh view, called "beginner's mind" in Zen, is the opposite of "psychosclerosis," the hardening of an attitude that cements our habitual modes in place.[1]

Modes are by definition transient—even if they drag on for hours or days or years, they at least have a potential end. Because modes are learned responses, they can be changed by new learning. But while we can be put into a mode by subtle cues, disengaging from a negative mode takes intentional effort.

Try this: cross your arms, but with the arm that usually goes on the bottom on the top this time. Feels a bit strange, right? That's what it can feel like at first when we try to change emotional habits; it seems awkward and unfamiliar. But as we repeat the new response over and over, the awkwardness fades and the new response gets familiar. With sustained effort, it becomes the new automatic routine.

Our comfort zones—the easy habits we've fallen into—make us complacent rather than prompt us to go beyond. At first, forming a habit takes brain energy, and resisting the seductive pull of automatic routines takes effort. So we need to challenge our modes by mindful awareness coupled with a resolve of willpower to oppose the old routines and start new ones.

While replacing a habitual routine with a new one takes awareness and effort, the more we repeat the change, the more the basal ganglia take over the routine's operation and the less brain energy it takes, since the routine then becomes our new automatic response.

When we intentionally shift out of a negative mode, we take charge of our own minds. Mode shifters open us to a mental clarity that transmutes negative modes and allows positive ways of being to shine through.

Both Eastern and Western models of dysfunctional modes recognize the futility of the deep habits we re-enact over and over in our lives and the suffering this causes. Some habits, of course, are more challenging to change than others. Our negative modes serve deep emotional purposes and so do not follow simple logic. In part, we have built our identities on a foundation of such modes; when we start to unhook them, we can feel a bit destabilized. That shift can bring up strong feelings, even doubts about making the change at all: If we are not these emotional habits, who will we be?

Our familiar modes define us to some extent, even though they may be outdated responses that no longer serve a useful purpose. It can be unsettling when they start to dislodge. We may feel a poignant compassion as we realize that for so long we've believed something distorted about ourselves or believed other people to be true. But when we see through the illusion, it no longer has the same power over us.

From an Eastern perspective, this shift in awareness is like using the sky as our reference point rather than defining ourselves by a cloud of unknowing. We still see the cloud but also see its transitory, transparent nature, and know that it will eventually evaporate as it melts in the warmth of awareness.

As we begin to free ourselves from modes that have confined us, we feel a sense of disenchantment with repeating relentless emotional habits. As we free ourselves from cognitive and emotional habits that have obscured our genuine nature, our awakened mind naturally flourishes.

Our Inner Stick Shift

Once Mahatma Gandhi was visited by a mother who was worried that her son was eating too much sugar. She asked Gandhi to tell her son to stop.

"Come back in two weeks," Gandhi said.

When they returned, Gandhi immediately told her son to stop eating sugar.

Why the delay? "I was still eating sugar myself the first time you came," he told her.

Gandhi was a firm believer in habit change. He once said, "As human beings, our greatness lies not so much in being able to remake the world as in being able to remake ourselves."

Mode triggers come to us unbidden, thrusting us into whatever mode they happen to prime. But shifters are ways we remake ourselves to build more positive habits. Without shifters, we're at the mercy of whatever triggers come our way.

Like modes themselves, shifters take many forms. Sometimes a shift is dramatic and sudden. Other times mode shifting can be more subtle; just a small nudge can create some breathing space or a few soft words to ourselves may open us to a wider perspective and a change in attitude. And with our most entrenched and emotionally charged modes, the shift can require dedication and diligence.

Our days are full of mode-triggering moments. But they come randomly, and the modes we are put in or plucked from are just as random. Why not take the helm ourselves and pilot our way to better modes?

Oddly, some people purposely practice putting themselves into bad modes. Bill collectors, I once heard, work themselves up into an angry, punitive mode to help them be more forceful with the people they are demanding payment from. Bill collectors seem the odd exception—most of us rarely, if ever, would choose to voluntarily enter a toxic mode. I wonder if bill collectors can turn off the anger at the end of a workday. Do they know how to shift back into a better mode? If not, how could they stand their jobs?

When we intentionally shift out of a negative mode or make any change in mode habits, we are taking charge of our own brain functions. In theory, the brain rewires a tiny bit each time we make a mode shift. The more we do it, the stronger that new wiring becomes and the easier the mode change gets—a principle known as neuroplasticity.

Neurons that fire together wire together. And we've repeated our less desirable modes thousands of times. But mode work re-educates the brain: if we can be consistent in making a mode shift, it becomes a well-practiced neural pathway. Repeated practice rewires the brain. As the basal ganglia network takes over the new habit, that response becomes the brain's default option—what we do automatically when the right cues come along.

Rebooting

For a while after her severe auto accident, Robin showed some of the normal signs of post-trauma stress. For instance, she would become intensely worried that an approaching car would hit hers head-on, particularly when she had to cross a bridge.

But Robin tackled this directly: she returned to Arizona, visited the people who had helped her, and walked the stretch of road where she had been in that head-on collision. "It was a rough five days, but so worth it," Robin told me. "Seeing where the accident happened, seeing that road. We just used the road like people do every day, and I realized that that location is not a place that hurt me, and it's not always going to be what I picture in my mind."

With traffic going along as normal, Robin saw the place with new eyes—as it actually was rather than how she had thought of it. "I could see that it's just a place that happened to be where this tragedy struck."

That, in turn, changed her attitude toward her own trauma. "It's not something that I'm going to carry with me. I was able to put a new spin on it, and take it in for the better."

She added, "They say that if you avoid thinking about the accident, it just gets deeper and deeper and you don't want to talk about it. But as negative as the whole experience was, I'm always willing to talk about it, and it doesn't feel like a huge burden. It feels more like a life experience that I can share with others for inspiration and show them that you can have something like this happen and you can grow from that."

Robin's visit to the scene of her accident will be a familiar moment to many people who have gone through therapy for post-traumatic stress disorder (PTSD) involving understandably upsetting memories.[2] One of the standard elements of treatment involves mentally revisiting the trauma itself, but from a secure mode. The memories are reappraised while feeling a sense of emotional security and so become stored in a less upsetting fashion.

Revisiting the scene of the trauma in person (or the memory of it with a person you feel safe with), while trying to relax as much as possible, actively replaces the fearful trauma-mode reactions with positive reactions. Eventually cues that reminded you of the trauma no longer trigger the full-blown reaction.

Each time we are able to re-appraise the events and memories that trigger a traumatic reaction, the memory traces for that cause-and-effect pattern are reconsolidated a bit differently at the biochemical level.[3] So, as we remodel our mode patterns, we are gradually rewiring our brains. With each successive reboot, the triggers, hopefully, become less potent and our thinking less distorted.

So it is with other negative modes too. Even a small shift in a mode routine can make a big difference. It may be something as simple as taking three deep breaths before reacting to a child's misbehavior, to head off yelling from a punitive mode. Or, in attachment mode, realizing in response to an impulse, *I don't really need this*. Or in the duty-bound mode, *I don't really have to do this*.

Mitate is a Japanese term for re-perceiving things, seeing in a fresh light. Though nothing necessarily has changed in our circumstances outwardly, our old reactions no longer define us as our perceptions begin to shift. More breathing space allows room for alternatives.

The Choice Point

Diana had just spent a wonderful weekend with a close friend in upstate New York and was driving back to Manhattan when she decided to stop at one of her favorite country farmers' markets on the way. The vegetables, exceptionally fresh, cost a tenth of what she would pay in the city.

While Diana was at the market, she thought of one neighbor after another and kept finding things she thought each of them would like—the elderly woman, the gay couple, even the neighbor who drove her a bit crazy. She felt such joy to be thinking of them and buying things that would make them happy.

So Diana piled the bags of farm-fresh food into her car and drove back to the city. Because she had so many bags to carry, she parked temporarily in a no-parking spot right in front of her apartment building. She enthusiastically carried several bags up to her apartment, making trip after trip. But she had forgotten to put her blinkers on, and on one of the trips down she found a parking ticket on her windshield.

"I sank into my *how unfair* mode, getting grumpy and sad," she later told me.

But after a few minutes of that deepening mode, Diana told herself, *Whoa!* as she realized what was happening to her. She reminded herself,

I just had a beautiful weekend and felt so much joy while I was finding those great vegetables for my neighbors. I'm not going to let that joy be ruined by these negative thoughts.

And with that came a heartfelt shift in her response: the joy of making others happy replaced feeling sorry for herself. She snapped herself out of that mode and substituted a better response; thinking of helping others primes the secure mode. She found a legal parking spot for her car and then continued delivering the vegetables to neighbors, all of whom were delighted. "Even my neighbor who drives me nuts was almost moved to tears!"

The crucial choice we have in every moment is to feed the distressing modes we fall into or to make an intentional shift into a more nourishing way of being. "When my toxic modes shout at me," as one student said, "mind whispering feels so gentle—a soft, still voice that can stop your craziness."

As we gradually re-educate our emotional habits, we no longer see through the distorted lenses of our modes. The road that forks in every moment leads us either to a negative mode or toward a way of being where we can blossom and connect from our own secure base in authenticity.

We can reach toward the light of wisdom or plunge into the darkness of unknowing. This universal choice point has been noted throughout human history; it is central in the great moral dichotomies of myth, religion, and ethics. In the religions of India, the choice has been put in terms of delusion or insight. For Socrates it was a life of obliviousness versus self-inquiry. It resonates with what in that Native American allegory boils down to the choice between our "two wolves"—resentment, envy, jealousy, greed, arrogance, and self-pity on one side; wisdom, kindness, empathy, joy, peace, and compassion on the other.

Likewise, the various sciences that mind whispering draws from each has its own lens on the choice, whether called "adaptive and maladaptive ways of being," "secure and insecure modes," "functional and dysfunctional brain states," or "adjustment and pathology." In every moment we can go down the path of a positive or a dysfunctional mode. One choice leads to fear or denial and anxiety, the other toward a more nourishing mode of being.

The choice lives on in mythic worlds like *Star Wars*, with its dark and light sides of the Force.[4] Think of the central character of the saga, Anakin Skywalker. As a boy, he endured emotional wounds—particularly burning anger at the tragic loss of his mother. This made him susceptible to being cast into a bitter, enraged mode later in life by the death of his wife.

Early in his life as a Jedi, Anakin Skywalker had embraced one mode of being, and after his wife's death quite another. His anger and bitterness made him ripe for seduction by the "dark side of the Force" into becoming the Sith lord Darth Vader. As one Jedi put the difference between these two modes of being, "The Sith thinks only of himself; the Jedi only of others."

The saga famously culminates with Anakin's redemption and return to the altruistic Jedi mode with the aid of his son Luke Skywalker. By rescuing his son from certain death at the hands of the evil emperor—and getting fatally wounded himself—Anakin breaks free of the dark mode.

At that moment Luke says to him, "Father, I have to save you."

Vader—now in the Jedi thinking-of-others mode once again—replies, "Son, you already have."

The Art of Whispering

There's a striking similarity between Bob Sadowski's approach to relating to horses and the way two people in a secure base together connect with each other. Bob, essentially, joins with a horse in the secure mode and works in a whole other model of learning, going well beyond typical "training."

Bob has been helping me and my horse learn in a spirit of play, even when we're doing serious work. Working with Bob one day, I was trying to be aware of these skills while sustaining a heart connection with Sandhi as we danced through movements together. Rather than me forcefully controlling her, we were working as one, enabling it all to unfold naturally.

I was learning new ways of moving the training ropes, the special ultralight leads, and the halter she wears during these sessions. To a horse, movement itself is a language. Bob reminded me to allow it to be simple: to see the essential moves more clearly and let the extra, unnecessary movements drop away.

I held one end of a thin training rope, which was attached to her as a halter, and I sent her light movements through that rope. The movements were meant to signal her to pivot in one direction or another.

Bob encouraged me to stay connected to her throughout, to mentally attune to her movements and emotions so that the process would be smoother, and to enrich our relationship with each other. I set an intention to do so. Then I gently positioned her and myself. At a particularly challenging moment in the process, when we were effortlessly gliding through the movement together and executing it seamlessly, I was nothing short of elated.

But suddenly Sandhi and I fell out of sync. She stopped, looking perplexed.

Just then, Bob reminded me not to let my feelings get in the way. And I saw how even a moment of excitement could subtly distract and interrupt the flow. My exuberance had created a confusing cue for Sandhi as she had been learning this new habit.

I tried it again, remembering the instructions, repositioning myself, and reconnecting with Sandhi, while again setting my intention, guiding and allowing, and then letting it happen. I saw the need to be simple and clear in what I was asking Sandhi to try and, if we weren't connecting, to re-assess what to do next.

This time there was an improvement: she read my intentions without getting confused. Finally, by the end of our session, she had performed exquisitely. Her reward was obvious: she had enjoyed learning and pleasing us. In fact, she seemed to shine with confidence.

Both horse whispering and mind whispering attune to what's needed in the moment by listening—in the case of mind whispering, listening to faint murmurs in the mind. This takes mindfulness. The Tibetan words that, together, are often translated into English as "mindfulness" are *drenpa* (literally, remembering to be aware and recollect the pertinent instructions) and *sesshin* (knowingness or consciousness, reconnecting with those instructions—in this case, the principals of habit change in mind whispering).

Both qualities work together in a sustained awareness that allows an open investigation of our modes in the moment and brings to mind a needed remedy. In traditional meditation methods, what's needed might be more energy and effort to counter laxity or more calm as an antidote to agitation. In mind whispering, what's needed depends on what mode we find ourselves in and how we are reacting. This means recalling the habit shifters for a given mode: we might need, for instance, more engagement and presence or more assertiveness.

A third quality is *bayu:* applying that choice wisely. Choosing what to adopt and what to avoid is the essence of habit change. This allows us not just to discern what's needed but also to put it into practice.

In mind whispering, *drenpa* is called "remembering," which includes a deliberate pause to remember the instructions. *Drenpa* pauses, checks the mode temperature, and asks *what's happening?*

Sesshin refers to mindful presence, reconnecting with those instructions to put them into practice. *Sesshin* takes the information from *drenpa* and asks *what's needed?* This gets us ready to put the instructions into practice.

Bayu is choosing how to wisely implement what's needed or how to apply a remedy in a particular situation. The three qualities act together as mindfulness instruction, adapting classic meditation into an applied, everyday wisdom for mind whispering.

These steps are echoed at construction sites in a sign urging workers to help them use safety precautions. It's a stoplight. The red light features the word STOP, the yellow, THINK, and the green, ACT. That sign serves as a reminder not to go through the motions of work on automatic pilot, but instead to take a pause, discern what's needed, and put that into action. Practicing these three steps together builds the habit of waking up to the present moment, finding the apt mode shifter, and changing the conditioned pattern.

In its original context in Eastern psychologies, meditation is a tool used in the most fundamental habit changes: from confusion to clarity, delusion to wisdom. With our emotional habits, it may feel as if we have little choice, given the layers of unexamined, unconscious behavioral conditioning that usually make up our modes. But our ability to be mindful about habit change—to make a wise choice to replace a knee-jerk impulse—enhances its effectiveness.

We may cycle through many modes on a given day, riding an inner roller coaster without realizing that those ups and downs track our modes. Mode shifting starts with the realization that we are indeed caught in a way of being that limits us.

At best, that recognition can be very specific, as long as we become familiar with the taste of our default negative mode—the typical thoughts, feelings, and reactions. But if it is not clear exactly what mode has been activated, for our purposes we can simply think of it as "insecure" or "distorted" and track our shifts toward the secure mode.

A Wise Choice

A physician from India, who works in a European country, had been having difficulty with a coworker who was very opinionated and always thought he was right. The physician had suppressed her reactions and felt subjugated in the relationship, as well as unable to say what she felt, which had triggered her avoidant mode as a way to cope. As a result, she was pulled out of the present and caught up in a distorted view of herself and the other doctor.

She would remain silent as he regaled her with his strong opinions and judgments about her work and that of others. Rather than reacting outwardly, she practiced taking a mindful pause and remembered to attune to her mode feelings. She recognized that her avoidant and prey-like modes had taken over and reminded herself about the mind whispering approaches she had learned.

From this more open space, she chose wisely by devising a response to her domineering colleague. Rather than avoiding the doctor or passively resigning herself to go along, she made a conscious choice to shift from her avoidant and prey-like modes to an assertive stance.

The next time he approached her in the same old way—in the past, a mode trigger for her—she responded in a new way, one that felt far better to her. She told him, "We each see things in our own ways. I have a different point of view than you do. It's okay we don't have the same opinions. I respect your opinion, but I don't see it the same."

She said this clearly, calmly, and firmly. This change in her response led to a new one in the other doctor too: instead of acting from a predator-like mode, he quietly walked away.

Amazingly, he came back the next day and apologized for being so dogmatic about his own opinions. He said that he knew this was a problem of his, and she had helped him see it in action. He actually thanked her. By altering her habitual mode reaction, she had opened the way for him to change too.

To unpack this even more, the physician had applied many practices and perspectives from mind whispering. She

- took a mindful pause rather than going down her automatic, habitual mode track;
- recognized a trigger: being subjugated;
- interrupted her usual coping modes: avoidant and prey-like;
- paused to recollect and remember the instructions and to attune to her feelings;
- discerned with insight her activated mode feelings;
- clarified what response was needed to make an intentional shift away from her habitual reactions;
- gave herself some breathing space from the mode's conditioning, to make a wise choice; and
- asserted her views with equanimity, replacing the automatic reaction with a more constructive alternative.

Her authentic communication signaled she was acting from her secure mode. When the other doctor apologized, his sincerity signaled the same. When we're in the secure mode, we can then more easily speak from the heart in ways that can touch someone else.

As we work with our modes and get more familiar with them, we more readily recognize telltale signs—those resentful thoughts, that yucky feeling again—of our most common negative modes and what to do about them. This is a crucial first step in beginning to change our mode reactions.

By bringing an automatic habit into awareness, we shift control from the basal ganglia to the executive circuits of the prefrontal cortex. Then we gain choice over what had been a mandatory routine. This shift lies at the heart of all habit change.

By putting some space between our awareness and our turbulent feelings we are more able to catch ourselves, even as we are about to go through the same familiar routine. We can step back and ask, *Do I want to make this real?*

Opening up a space in our minds gives us more choice in the moment, so we can recognize that a habitual mode has been triggered. Then we have the chance to make an intentional shift.

The hidden ingredient in this transformation is awareness. A mindful pause tracks the cues that trigger our modes—or recognizes the mode itself—and lets us access the remedial shift. And every time we free ourselves from the grip of habit, that very act is a small liberation, a reward in itself.

A Calming Pause

Diana Broderick, a Manhattan-based interior decorator, had finished a job at a client's house. Some time afterward she realized a stepladder had been left there.

Several weeks later she got the client's okay to go back to the house for the ladder. As she unlocked the door, Diana heard an alarm go off, and she realized she had forgotten the code. She found herself frightened, not knowing what to do, and terrified that she would be arrested when the police arrived. When she again tried to remember the code, her mind went blank.

She frantically called the client's cell phone but got no answer. It seemed there was nothing she could do but wait for the police and hope that she could explain her presence to their satisfaction.

Diana took several deep breaths and then recognized the anxious mode that had been rapidly capturing her mind. So she decided to try a calming meditation she had learned, fully attending to her breath to focus a bit, then turning her mindful attention to the flow of her thoughts. As she started to watch her mind—instead of being swept away by the torrent of thought—Diana found herself quieting down and relaxing, despite the clang of the alarm. As her mind became more peaceful and open, the alarm code resurfaced in her memory. She leapt up and punched in the code, before the police got there.

Mindfulness of breathing calms us and is a direct antidote to anxiety in any mode. Diana's mindful pause had allowed her to reconnect with awareness (*drenpa*), which knows what is needed (the insight of *sesshin*). Her wise choice (*bayu*) had allowed her to observe her breath. When she'd been able to calm her anxiety, a lightbulb had lit in her mind. Sometimes the illuminating insight we need is just an alarm code!

Such in-the-moment mindfulness offers a powerful tool, whether to help us think more clearly or to weaken the hold of even our worst modes. As we become mindful, we change our relationship to our own minds. Instead of being trapped in these mental patterns, we can begin to *see* them.

Pausing mindfully to become aware of whatever pattern is active can take the form of tuning in to our emotions or our intentions, or simply aiming attention at whatever is most present in awareness. Getting into the habit of examining our minds this way lets us interrupt the learned routines that lock us into our modes.

A mindful pause lets us step back from immersion in the stream of our experiences and be aware of what's going on in and around us. Then, if we find ourselves in the grip of an unpleasant mode, recollecting the choices we have for remedies lets us choose what's needed.

From awareness we are more able to recollect what to do, so we can apply what's needed. This step alone can sometimes be enough to break the spell of a mode and return us with more clarity to the present.

In a famous study, neurosurgeon Benjamin Libet found that there is roughly a quarter second between our becoming aware of an intention, like to move a finger, and the actual movement of the finger. During that gap we are aware of the intention, but we can choose not to move. Some call this the "magic quarter second"; I call it the *mindful* quarter second.

These reflective pauses represent an invaluable awareness tool in mode mindfulness and give us greater choice in how we respond. Brief mindful pauses throughout the day let us note our intentions and aim our attention at what is needed in the present moment.

Pauses with mode checks can reveal the intentions behind actions you are about to take, giving you greater choice. Negative intentions often reflect dysfunctional modes. Positive intentions, like acts of kindness, are typical in the secure mode or with a wise heart. We can make the wish to have positive intentions, which leads us to positive choices.

Mindful pauses can develop into the habit of checking our intentions and distinguishing between our choices as well as aligning our choices more with what's needed in the moment. You can make this practice your own by playing with the ways to pause that feel most relevant and natural for you.

For instance, you can integrate this practice with closely observing intention during walking meditation. In mindful walking you pay careful attention to the sensations of movement in the legs and feet as you walk slowly in one direction for ten to twenty steps. After you mindfully walk to the end point of your path, pause and closely observe the fleeting intention to turn before you begin turning around to walk in the other direction. Then make the turn and walk back. You're not trying to get anywhere, but simply trying to pay full attention.

While mindfulness of walking may seem to have little to do with mode shifts, it cultivates the more general capacity to attune ourselves to what's going on while we are being active. That's essential for mindful mode change.

Getting into the habit of sprinkling the day with mindful pauses lets us sense when the modes are taking hold. Mindfulness acts like a mental immune system, roaming its domain to repel undesirable invaders. We begin to deactivate a negative mode the moment we become aware of it. As mindfulness gains momentum, we can better neutralize unhelpful modes as they start to take hold.

What Does This Mode Need?

Once I was driving down a road on the outskirts of town when I noticed that two unfamiliar horses had gotten out of their pasture near a wooded area and were grazing in the wide yard of a house. I pulled over and jumped out of the car, then removed the belt from my jeans to use as a lead rope to catch the horses. If I caught one, I thought, probably the other would follow along. Then I would tell the people in the house that their horses had gotten out of their paddock. I was happy to be fulfilling my responsibility as part of the neighborhood watch in my town.

As I ran up to the horse and got ready to throw my belt around its neck, I suddenly saw antlers. I realized that it wasn't a horse at all but a moose!

My mind stopped. As my car's GPS says when I make a wrong turn, "Recalculating."

The moose looked just as perplexed as I felt. They had probably come out of the woods in search of greener pastures. Using bear tactics, I backed away slowly, thinking, *Nice moose,* and hopped in my car.

"Recalculating" can be a helpful reminder in other situations, besides trying to catch a moose. It can be useful in catching us before we get hooked, reminding us, *Don't go there!* Instead of just doing the first thing that occurs to us as an automatic mode pattern plays out, we can rethink that habitual response and replace it with a more appropriate one.

Recalculating also represents the playful spirit we can bring to mode work; playfulness helps in awakening us at every level. Say you're enthusiastic about going for a hike in a park, and you approach a friend about it, assuming she will love the idea. But your friend gets a bit irritated that you didn't first check to see what she was in the mood for, which happens to be visiting a new exhibit at a museum. Giving yourself a lighthearted, playful reminder—recalculate!—can be enough to marshal a flexible awareness and shift gears. Perhaps a visit to the exhibit plus a vigorous walk in a nearby park would be a happy compromise.

Once we recognize when we are in a distorted mode, then we can take steps to clarify our thinking and feelings and act from a more skillful place. So the first step is to bring the mode into the clear light of awareness. The second is to apply the corrective antidote, a positive alternative.

Knowing the various remedies for a mode gives us a recalculation menu to choose an apt corrective. For instance, understanding that concentrating meditation practices have a calming effect, we can apply one to lessen anxiety when we need to. When it comes to altering a core belief, the remedy might come from cognitive therapy's methods for challenging assumptions and cognitive distortions.

Each mode has its own correctives. But there is no universal checklist of mode remedies that works for everyone in all situations. Each of us has a unique personal history, an individual way of inhabiting our modes, a specific set of triggers and core beliefs. We may find our own individualized remedies (more on these later). Educating our minds and hearts can lead to a growing confidence in our awareness and trust in our inner resources.

Every mode has elements that might be changed in specific ways that are reparative. Take the avoidant mode. The underlying need is to feel safe confronting distress and engaging with intense feelings. Any shift that helps you learn that strong negative emotions aren't always threatening will help you to feel comfortable with intimacy rather than needing to distance yourself.

Remedies for the anxious mode might include challenging exaggerated fears, learning how to stop worrying and calm down, being more self-contained, or getting comfortable with having more space in relationships and being less intensely connected. Or a remedy might challenge the assumption that distance in a relationship means you are losing the person altogether.

Such shifts offer some rough answers to the key question: *What does this mode need?*

The Habit of Habit Change

Mary Beth was talking about a friend who had disappointed her: "I was about to be judgmental, but I stopped it before I said anything. I like to give people the benefit of the doubt rather than spread negativity."

Where had she developed that habit? On a two-week mindfulness retreat she had simply noted whatever thought or feeling came up in her mind, without reacting to it or getting carried away in a train of successive thoughts. What came up for her repeatedly was a constant stream of judgments: about herself, other meditators, the teacher, her clothes, their clothes, her zafu, the lunchtime tofu—anything whatsoever that entered her awareness.

But with sustained effort she learned to stay even-minded, especially with the negative judgments of herself and others that she was so prone to. She saw that she did not have to embrace these thoughts; she could just let them go. Ever since that retreat it became a habit to apply this mindful radar in daily life, to catch herself when her mind turns judgmental.

You can play with changing habits; you don't have to tackle modes right off the bat. Even shifting a minor routine strengthens your confidence in being able to change. Every time you say no to a habitual action and substitute another, you build willpower. And you just might feel a bit more in control of your life as that habit of challenging habits spreads out to other routines.

Start with something small and easy to accomplish. Say you usually half-make your bed in the morning, carelessly throwing the covers on. Resolve to make it more carefully, leaving it perfectly neat. Applying that positive intention will require a moment of mindfulness as you get up in the morning, so you remember to change the routine.

Once you've done it the new way for several days, you might add a more sustained mindfulness. Slow down a bit to be aware of all the movements involved: bending . . . reaching . . . lifting . . . placing . . . tucking. Try not to let your mind wander to other things, but rather stay closely aware of all your movements. Make an effort to be more present than absent with what you are doing. That alone begins to shift the habit of halfhearted to wholehearted attention.

Notice how this makes you feel. Do you sense a small reward from being more organized?

At first it may seem a bit boring to make your bed mindfully. After all, free time like this is usually a good opportunity to review your list for the day or at least plan what you'll make for breakfast. But if you use these daily, typically mindless habits instead as a chance to practice being fully present, you're putting in time strengthening the mind's overseer and increasing your capacity for presence.

As you change the habit of being distracted during ordinary routines to using them to sharpen awareness, it enhances your general capacity to spot and challenge your mode ruts. Each time you notice a habitual reaction and make a change for the better, you strengthen neural pathways that support the new response. Plus, you send a message to your brain that will make you more likely to tackle more entrenched mode habits.

Our maladaptive modes create a sort of oppression in the mind.

While we are in their grip, we often fail to sense how they imprison us—something that becomes apparent the moment we investigate them with a mindful discernment. These modes act as guards for the mind, spinning around on the spot to check what's come our way, understand it, render it harmless, and let it go.

With awareness alone, we may see that we are caught in our modes, but they can still continue to play out. By adding discernment and a wise choice, we are more likely to find ways to interrupt their compulsive agendas.

I Am Not My Modes

A social worker was sent with an aid team to an African country to work in a refugee camp just after a bloody civil war had ended. She had been physically abused in childhood. Many of the people at the refugee camp had been tortured or abused, and the social worker would freeze when she encountered a refugee whose plight brought back memories of her own abuse.

She had been in one of my workshops and had taken her notes along on this trip. At night she would read them under her covers (so as not to disturb the other aid workers she bunked with) to remind herself that these overwhelming feelings—as the social worker later said—"were just patterns of habitual emotional reactions, not who I really am."

That may be the fundamental perspective to bring to our modes: they are like passing clouds. Such corrective learning, when brought to mind while we are in the grip of a mode, helps change our relationship to it and loosen its grip.

This crucial step takes a mindful pause; we tune in to the present. Mindful presence can allow our wisdom faculty to arise, which is the ability to tap into what cognitive scientists call the "adaptive unconscious," the huge part of the mind that stores much of our understanding in ways often not accessible to the part that puts thoughts into words.

We get a sense of what has gone wrong. Then, tuning in and discerning what we need can reveal the right remedy, approach, perspective, or practice—or just help us be present with full awareness of whatever occurs.

The first step applies discernment in order to become familiar with what your main negative modes feel like: what they compel you to think, feel, say, and do. With sustained investigation, we build a mental capacity that empowers us to spot and counter these automatic thoughts and the underlying beliefs along with the actions and feelings that flow from them.

As qualities of mindfulness enhance our awareness, we can allow this precise, open, clear awareness to shed light on our emotional patterns so we can respond more skillfully when our maladaptive modes arise. And to the degree we sustain ourselves in the secure base of mindfulness, we are more immune to the pull of our negative modes.

CHAPTER

10

The Mindful Overseer

On a huge soundstage in London, the famed composer and conductor John Williams was leading the London Symphony Orchestra in a rehearsal of a movie soundtrack he had written. Sitting in the audience watching these world-class musicians play their parts, each with utter precision, attentiveness, and confidence, I was fascinated by how each contributed their portion of the whole piece right on cue.

At the center of it all stood Williams, the conductor, like a regal leader of his musical herd, seeming to pay keen attention to each musician individually while leading the orchestra as a whole into a flawless arrangement of melodious sound. Even this rehearsal sounded like a polished, extraordinary performance.

It was fascinating to see how the conductor coordinated each section of musicians: strings, percussion, horns—spotlighting this soloist, then calling on a group of instruments, then gathering the whole orchestra—like a dancer with ninety-six partners. Somehow he seemed both invisible yet majestic, knowing when to be still and when to guide, all in a graceful flow. He'd lower his hand to reduce the volume then raise his arms to build the sound into a crescendo.

As I watched the conductor listen, attune, and execute, it struck me that his performance illuminated a powerful truth: we all have our own inner conductor playing a parallel role as our minds' overseer, orchestrating the multifarious voices in our heads. Our minds' overseer likewise listens, attunes, knows what is needed, and guides us to act accordingly. It is the driver of wise choice.

The area that houses the brain's conductor lies just behind the forehead in the prefrontal zone of the neocortex (the thin sheaf of layers covering the top of the brain). The prefrontal cortex has more connections to other parts of the brain than does most any other neural zone, giving it a unique aerial overview—like John Williams up on his podium. This wiring lets the prefrontal cortex act as the brain's chief executive or manager, something like the conductor of the thousands of instruments in the brain's orchestra.

Our sense of purpose, our goals and motivations, our abilities to plan and think creatively in the face of life's adventures and challenges all depend on the prefrontal zones. This area of the brain coordinates endless operations, from computations and generating new ideas to texting or cooking a soufflé.

In addition, a vast medley of incoming stimuli—sights, sounds, tastes, smells, touches, and thoughts—needs to be managed so that we only deal with what matters now instead of being overwhelmed. To keep our mental house in order, the prefrontal cortex is like Williams, enabling us to get smooth harmonies rather than an endless cacophony.

Fittingly the jobs done here are called "executive function"; the prefrontal area leads the rest of the brain. Multiple circuits within the prefrontal cortex operate as the mind's manager.[1] Together these create an executive committee of the mind. As in any highly effective group, the committee "members" each bring unique strengths and abilities:

- The Attender directs our attention, determining what we notice and what we do not. Function: Attention.
- The Perceiver refines our sense of the world by detecting patterns and meanings in what we notice. Function: Appraisal.
- The Verbalizer puts what we notice into words and tells us the sto-

ries that help us comprehend what we perceive. Function: Understanding.

- The Motivator attaches emotions to what we perceive and so makes some things more important than others, determining what we care about. Function: Drive.
- The Coordinator puts together our perceptions, understanding, and feelings, decides what to do, and then guides our actions. Functions: Decision and Execution.

In short, the prefrontal circuitry directs attention, appraises what we notice, articulates our understanding, compels us to act, decides what to do, and directs the resulting activity. The key steps in mind whispering—*drenpa, sesshin,* and *bayu*—roll together all these executive functions. Mind whispering integrates mindfulness with habit change, which adds the strength of awareness to the mechanics of shifting modes for the better.

Our habits, remember, are automatic routines orchestrated by our basal ganglia, the primitive network we share even with reptiles. By activating our prefrontal area, we dredge up these long-buried habits and bring them into the light of attention, where we can reappraise them, understand their deficiencies, determine changes, and put that determination into action.

What may matter most for mode shifting is the prefrontal zone's power, as one neuroscientist puts it, as "our mechanism for liberating ourselves from the past and finding our way to a better future."[2]

The Mind's Manager

My horse, Sandhi, was learning how to walk through a narrow path bounded by electrified wires. A wrong move meant a quick, sharp jolt—no physical harm, but something dreaded by horses.

Bob told Sandhi what they were about to do as he held the lead rope. He took some steps with her walking beside him, then he stopped for a pause—and she halted too. They did this repeatedly as they went down the path.

When she's not in her secure mode, Sandhi's mode of choice seems to be the horse equivalent of the anxious, clinging mode. As Bob and Sandhi got closer and closer to the meadow—and as Sandhi got stronger whiffs of the grass below—she became visibly excited.

By the time they reached the field at the end of that electrified path, Sandhi's eyes bulged with longing and her body brimmed with energy. She was clearly eager to curl her lips around those juicy leaves of clover and blades of gorgeous green grass.

But Bob said—rather signaled—to Sandhi, *I'd like you to stay still until I unhook your lead rope and I tell you it's okay to go.*

If it had been me holding that rope instead of Bob, Sandhi would have been dragging me behind her 800-pound body. I was amazed to see her patiently standing still beside him, as though impervious to the luscious field of greens.

I was even more amazed at what happened when Bob unhooked her lead and with a gesture told her calmly that she was now free to go and graze. Sandhi just continued to stay right there as if glued to his side. It seemed she felt being there with Bob was more important than grazing. Finally he urged her to go by making a more emphatic gesture, and she plunged ahead to graze, as though savoring a well-earned reward.

This lesson applies to us as well. It reminds me of the "marshmallow test," an experiment performed at Stanford University. Four-year-olds were seated, one by one, at a little table where a marshmallow was displayed. They were each told that they could either have the marshmallow right away or, if they waited a few minutes, they could have two.

After hearing about this test of impulse control, a workshop participant told me, "When I feel those urges to check online while I'm in the middle of something else, I tell myself, *Don't go online now. You can get online twice as long in a few hours!*"

This tussle between prudence and impulse is built into our brains. The limbic brain drives impulsivity (*Buy now, pay later!*); these neural circuits respond to immediate rewards and ignore later consequences. The prefrontal area governs discernment and considers the long-term, not just instant pleasure.

The neural root of the attachment mode can be traced to the circuitry for reward, which enjoys guilty pleasures like watching a string of cute-pet YouTube videos instead of doing work. It's the prefrontal circuitry that weighs in with a slower (in brain time) but more considered evaluation of such temptations. When we stop to bring a discerning awareness to bear, weighing all options with deliberation, we act from the prefrontal circuitry and sway the interchange between limbic impulse and cortical thoughtfulness in favor of a wiser choice.[3]

The pause that delays gratification—like my horse waiting to graze that luscious meadow—reflects the executive function in operation. Sandhi's stops on the way to the field of grass were a "marshmallow test" for horses. Bob was training her in restraining her impulse to grab for grass as well as keeping her safe from the electric fence along the path.

Just as Bob restrained Sandhi's impulses, our minds' executive can potentially override or veto the operations of an unhealthy mode when they conflict with our sense of how we want to be and live. What was once dictated by the knee-jerk responses of a negative mode can be replaced by a better choice.

When we can let a mode habit spend its force without our acting on it, we have weakened the hold it had over us. "Refuse to express a passion and it dies," wrote the pioneering psychologist William James. "Count to ten before venting your anger and its occasion seems ridiculous."

Creating a gap between impulse and action—which is precisely what Bob was doing with Sandhi—strengthens the mind's self-governing abilities and makes good use of that mindful quarter second.

Discernment

I heard there was going to be a gathering in another city of a group of close friends whom I had been missing lately. When I realized my work deadlines would keep me from going, I felt disappointed; I had been feeling isolated from my friends while working so hard.

Then one of those friends called and encouraged me to come. I thought it might be a good idea to take a day off, and I started to set things in motion. I was looking forward to it. I imagined reconnecting with good friends and projected onto my mental screen the pleasure of good company. But when I paused these Coming Attractions to check the weather, I read that a bad storm was coming and started to drag my feet.

I had been in the midst of researching and reflecting on interdependence and cause-and-effect sequences in the mind, which propel us into action. This ancient habit-change model, at the heart of Buddhist psychology, adds the power of mindfulness to mode shifting. When something seems pleasant, the mind projects a story of savoring a pleasurable experience. No problem, until things change and there's a rewrite in the script: I replayed the edited sequence, now imagining slippery roads, and I lost my momentum. I instead focused on work and, with some regret, reluctantly decided not to make the trip.

It was like rewinding my movie, going back to the bare facts without projecting my own liking and disliking. Later, as I gathered information about how things *actually* played out rather than how they appeared, I heard that most of my friends had also canceled. Having reframed the sequence with these bare facts added, I was feeling glad *not* to go—and I stayed home and wrote this instead.

In the passing show of life there are things as they appear and things as they are. One form *sesshin* can take is "wise discernment": investigating rather than simply accepting our mode's assumptions. This happens when we can catch the habit cue before it propels us into action. It lets us make a wiser choice.

Wise discernment mobilizes the critical faculty of insight. We distinguish things as they are by using a mind endowed with interest and openness, not clouded by judgments. Gathering more information— acknowledging that our projected inner movies do not always line up with the actuality—helps us make more discerning choices. There's a fine line between being discerning and judgmental. Discernment means weighing the pros and cons of different choices we might make and challenging our habitual mode responses when we see their drawbacks. With discernment our choices are in tune with the reality of a situation, translating our intentions into good results.

Take, for example, the negative modes typified by ruminating on worries. Brooding gets us nowhere—over and over we passively focus on repetitive mental loops of distorted thoughts, which makes us anxious and gloomy. But with discernment we constructively engage what's worrying us in order to resolve our difficulties. Instead of just cycling through the same set of worried thoughts, we go another step; we think through a range of responses and what their outcomes might be.

The worst impacts of ruminating are when someone feels both helpless to change things positively and has an intense yearning for things to be better. In one study, people who fell prey to brooding when they were feeling down or upset were found to be just as depressed a year later. But those who engaged in discerning reflection when they felt upset were much less depressed.[4]

In a study of "everyday wisdom," looking at what people learned from troubling experiences, the wiser ones were able to step outside themselves, reflect calmly, and see the crisis as a problem they could solve (or accept their predicament when they had no control).[5] Discernment helps us learn from challenges, so we're able to reflect calmly and see the situation from many angles—especially from another person's—and take action where we can.

Attunement and discernment work together, as helpers for the overseer. At the heart of attunement are empathy and compassion. And compassion builds a calm space in our hearts and minds that makes it easier to be discerning, just as discernment makes clearer the benefits of compassion.

The 17th Karmapa says that when he wants to know what to do, he stops and lets his mind rest in itself and "sees what moves and what knows." He encourages us to do the same—to get instructions from ourselves, from within our own "wisdom mind."

One classical benefit of mind training is in heightening our powers of intuition, allowing us to get a felt sense for a situation or person by tapping into our deepest sense of things. This offers a different way of knowing and is in contrast to more reasoned consideration.

Sometimes we have the insight we need somewhere within us. We just don't sustain our attention long enough for it to be revealed. We get distracted or often look outside ourselves for answers.

In a symbolic sense, wise discernment about our habits of thought and action, combined with a warmhearted attunement to our emotional patterns, is a kind of alchemical philosopher's stone—its inner force can reveal wisdom on remedying our modes.

Reappraisal

As someone wryly put it, "My mind is like a bad neighborhood. I try not to go there alone." But with mindfulness we are not alone. It offers us an unflinching, tough companion on this inner journey, one that lets us see things as they are rather than as they appear when we are locked in a negative mode.

On my satellite TV system a pop-up bubble occasionally flashes on the screen that says: ATTENTION! MODE REMINDER. The bubble reminds me how the remote control has been set, giving me the choice to change the setting. That seems an apt metaphor for how a given mode locks us into a narrowed set of options and how being mindful of our modes creates a powerful choice point.

For instance, a client told me about a time she was stroking her infant son's back and relishing the richness of the contented love she was feeling from this intimate connection. This went on for quite a while. Then she found herself taking a quick mindful pause to check in with her own mind. To her chagrin, this pause let her realize that for the last several minutes she had been plunged into an all-too-familiar troubled reverie of self-criticism, sucked in by her anxious mode again. "I realized," she told me later, "I wasn't even feeling my son's back but just going through the motions."

Mindful pauses let us sense when the modes are taking hold. Mindfulness acts like a mental immune system, roaming its domain to repel undesirable invaders. We begin to deactivate a negative mode the moment we become aware of it. As mindfulness gains momentum, we can neutralize such unhelpful modes as they start to take hold.

A woman doing mindfulness practice on a beach said later, "I felt lots of aversion. I didn't like the feeling of the seaweed on my feet when I walked into the surf. And there were rocks—I felt aversion to them too. But once I settled into being mindful, the aversion disappeared. I was able to see the extraordinary beauty of my surroundings."

Another tactic is standard in cognitive therapy: challenging our distorted beliefs and replacing them with a more realistic view. That woman could have questioned the thoughts driven by her aversive mode, examining whether they were supported by evidence, entertaining alternative explanations, and otherwise testing their logic. After all, there's nothing inherently negative about seaweed or rocks.

All of that takes the focus away from "my" reactions to explore the reality of what we're reacting to. This redeployment of attention away from the self is a method that Buddhist practice and cognitive therapy share. As we free ourselves from self-referencing, our negative modes deflate and our concerns become less compelling. Our mental energy gets freed and our hearts soften. We can be more understanding and empathic toward others, and replace resentment and worry with compassion.

Reappraisal—our second thoughts, not our initial reaction—holds a key. Life's annoyances can be sources of real angst or mere fodder for a good story, depending on how we perceive them. With these second thoughts, the brain begins to shift gears, activating areas in the prefrontal area.[6] Instead of just letting our minds take us over, we can become active shapers of how events impact us. While a negative mode fixates us in a rigid way of thinking, perceiving, and acting, the overseer allows flexibility—we can take in information fully, understand it from many angles, and respond as needed rather than in the hardened ruts of the mode.

In the grip of a negative mode, our thoughts and feelings rush by in a blink, giving us little choice over what we think, feel, or do. But when we activate the brain's overseer, we open the ways to more inner freedom of choice. The impressions of the steps we had taken down the path of confusion gradually disappear in the shifting sands.

The Mind's Overseer

One morning I woke up in a bad mood. Even small things rubbed me the wrong way: I ran out of tea, read some upsetting news, found I was left hanging on an urgent travel decision because I hadn't heard from someone with crucial information. I got impatient, frustrated, and grouchy. Everything I thought about seemed filtered through an aversive lens.

As I saw my mind starting to solidify around this grouchy mode, I thought, *I know where this day is headed.*

Then I thought, *Do I want to make this real?*

I saw how my inner little kid was about to have a meltdown, and I decided to refocus it to reboot the morning. I read an inspiring article about an ex-monk in Madrid who spent three decades building an entire cathedral on his own out of salvaged local materials. I watched a short video of my good friend Rose doing dazzling aerial flamenco, swinging on a rope high above the stage. With these prompts, I noticed an internal shift to a lighter state of being.

When we catch a negative mode starting to take hold, sometimes all our minds need is a gentle reminder that we are caught in a downward spiral. This kind of "remind-fulness" seems more and more important in these days of distracted multitasking. Amidst the inner clutter we can so easily slip into negative modes of being without noticing.

But when our awareness steps back to notice, we can begin to free ourselves. Often this attunement alone can be enough to release a negative mode's hold. Mindfulness activates the overseer, our inner guide, which can note our minds' malfunctions, recognize the causes, and apply an appropriate remedy.

There are many technical definitions of the term "mindfulness" within various traditions of Buddhism. We need to be clear about the difference between mindfulness in its original setting, as part of meditation training—respecting the traditional context and its authentic meaning—and how we adapt this faculty of mind for applications like mind whispering.

During meditation practice, mindfulness gets put to use in a very particular way: to steady the mind and sustain the connection to what we're aware of or doing. In daily life, mindfulness lets us notice when our minds have wandered and lets us return our focus to what's at hand.

In either case, mindfulness operates as the mind's overseer in spotting a problem and applying an appropriate remedy. It gives us the inner space from which to calibrate what we need in the moment, whether a heightened resolve or just a bringing back of our wandering attention.

Such meticulous attention hones the mind's ordinary capacity to focus. No matter what else goes on in our experiences, there is always an element of attention. Mindfulness sharpens that beam of awareness and aims it back on the mind itself.

This allows us to sense a finer level of subtlety than does our ordinary attention. We can look behind the veils in the mind that usually hide how the mind itself operates and bring them into the spotlight of attention.

This nonreactive focus steadily observes, without judgment or reaction, attending to whatever arises in the mind, neither choosing it nor turning away from it—a choiceless awareness of whatever is occurring in the moment.

Such equanimous observation—detached but receptive—avoids the extremes of suppression or reactivity, which can unfold into the modes of avoidance or anxious rumination. This noninterfering quality allows us to observe clearly how reactions build up in the mind and what might motivate them.

A mindful awareness can be seen as a mode shifter in itself. From the mode perspective, as we become more mindfully present, we are creating what amounts to an inner secure base—one that does not depend on someone else to help us but rather one we generate from within. We can become replete on our own, the more stable we become in this capacity of awareness.

Mind whispering builds on an underlying foundation that the Tibetan tradition calls "learning, reflection, and meditation." Take the first: learning. If you don't learn the correct view, for example, of one-pointed focus in meditation, then you can actually be training in distractedness. Or you could be enhancing clinging by thinking about a red sports car when you were intending to be paying full attention to your breath.

In mind whispering, learning includes becoming familiar with your mode dynamics—how the modes think, feel, act, and interact. Once we see a pattern of a strong emotional reaction and its underlying causes, our responses can be more apt.

Reflection means attuning and discerning: knowing with your heart your own inner feelings and empathizing with yourself and others, while also knowing clearly how your mind is working. When discernment combines with a compassionate motivation, the two work together. Compassion reduces our fears and distrust and opens a calm space in our hearts and minds, which enables discernment.

Ideally, attunement leads to lovingkindness and compassion while discernment leads to wisdom and a knowing mind, which illuminates things as they are rather than as they appear to be through our mode lenses. Compassion loosens the grip; insight releases it.

Applied Mindfulness

A Tibetan teacher explained the basic purpose of mindfulness this way: "We never know what the next moment is going to be. It's always something new. The function of being mindful is that whatever happens, you're aware of what is happening in that moment. There's no real objective to mindfulness beyond just being aware of whatever is."[7]

It helps to understand the purpose of a practice before we apply it.[8] So what is mindfulness for? It helps us manage the various states of mind it tracks and allows us to see what's needed in the moment.

Adopting an attitude of just being with things as they are is not enough for dealing with modes. While this gives us an essential, enhanced awareness of what's going on in our minds, this alone may not be sufficient for the changes we need.

In mind whispering we need the three steps: remembering to pay attention, considering remedies, and choosing wisely. This application borrows these steps for a different purpose than in Tibetan schools of meditation practice, where mindfulness is seen as playing a mental manager's role of maintaining all other spiritual practice we might be doing.

Mindfulness acts as a knowing awareness that observes the causes of problems, and an intuitive awareness that knows the apt solutions. As my vipassana teacher Sayadaw U Pandita says, "Mindfulness brings everything into balance on its own."

Some meditation texts describe the overseeing function of mindfulness in great detail. If our minds settle into torpor during a sitting, mindfulness alerts us, and so we can take action to correct the problem:

sit up straighter, widen our eyes, take a deep and intentional breath, get up and walk around, drink some tea—whatever will work, within the customs of a particular tradition.

If our concentration has evaporated into a medley of thoughts and fantasies, a mindful pause wakes us out of distraction and reminds us to focus. And if we've gotten too agitated and tense, mindfulness alerts us to apply ways to calm down and relax. Mindful awareness works as an overseer, checking our minds' operations, noting malfunctions, recognizing their causes, and applying the appropriate remedies.

A key distinction is between the function of mindfulness and the practices it manages, such as concentration or lovingkindness. Mindfulness sustains concentration, investigative inquiry, or compassion, but it is distinct. It manages each of them.[9]

So during meditation, mindfulness as the mind's overseer monitors how our practice is going. It reminds us to remember the instructions, reconnect with our awareness, reset our intentions, check if our practice is accomplishing what we intend, and execute all of this.

While we meditate, our mindful overseer checks on the quality of practice. We may hear a subtle inner voice saying, *Now you need more one-pointed focus or more insight, more equanimity or clarity, more compassion.*

Awareness is the fire in our emotional alchemy, melting the emotional clouds that obscure our inner nature. Awareness begins as we put effort into observing what we think, say, and do.

As we practice mode mindfulness, at first the effects of this insightful clarity may be fleeting, lasting only until the next emotional cloud forms. But rekindling this fire over and over is at the heart of this practice. As we learn to sustain mindfulness as a steady presence, the haze in our minds can clear.

Our minds' executive capacities can be reparative for healing our toxic modes.[10] For instance, the healthy adult mode acts as an overseer, monitoring our thoughts and feelings and replacing or modifying unhealthy patterns to help bring about healing. A remedy might result: say, to refuse to follow the too-demanding dictates of the duty-bound mode.

When we sustain an even attention to our inner experiences, recognizing our modes and their triggers without getting carried away or lost in them, we increase our chances that the mind's overseer will sense what we truly need at that moment.

As our abilities to listen to the whispers of the mind gain momentum, this mix of nonreactivity and penetrating insight energizes the force of our awareness, illuminating the path to further disentanglement.

Sustaining Attention

Remind yourself to be present and aware of whatever is occurring in the moment and to know the present moment's experience as it actually is rather than denying it or wishing it to be otherwise.

As sounds, sensations, thoughts, images, or emotions appear in the background of your awareness, allow them to remain there and focus primarily on the sensations of breathing. As these move into the foreground of your awareness, whatever predominates becomes the focus of your attention.

Sustain your attention without preference or indifference. If you find your attention wandering here or there, don't worry about it. Simply recollect your attention and stay with the natural flow of your breathing for a while.

Concentration can sustain focus and clear attention, allowing more room to accommodate experience without eliciting reaction, but with a tolerance, leaving you present to experience the moment as it is.

Allow a penetrating awareness to discover the true nature of what is occurring in the moment.

Be present and aware of whatever is occurring while connecting and sustaining your awareness.

Be aware of what's occurring in the five senses and the mental field.

When thoughts arise don't pay much attention to the them; pay attention to the awareness that recognizes them,

without lingering in thoughts of the past,

without entertaining thoughts of the future,

without conceptualizing the present.

The mind might feel cluttered, but there's still space:

space around thoughts,

space within thoughts.

The stronger our abilities to sustain attention, the more agile our minds' overseer becomes. The overseer can take the form of an ongoing awareness or a faint whisper in the background of the mind. Mindful presence naturally sharpens this capacity.

Mode Work

The Dalai Lama starts his day at about 3:30 every morning (he goes to bed at 7:00 P.M.!) with several hours of meditation—particularly a method that applies a discriminating inquiry into his thoughts. He is continually investigating, challenging assumptions, surfacing distortions, and reappraising, but he does it all with a profoundly compassionate heart. He exemplifies a revolution of the heart, bringing empathy and compassion to bear as well as an investigative mind.

Jeffrey Young and, indirectly, Aaron Beck have been cognitive therapy inspirations for me; both men are brilliant and have sensitive, intuitive hearts. But another profound model of investigating the mind has been His Holiness The Dalai Lama.

The Dalai Lama and Aaron Beck met to share their viewpoints in a dialogue at a world congress on cognitive therapy in Sweden. They connected easily and were genuinely interested in each other. During their conversation, Dr. Beck explained his approach to challenging distorted beliefs.[1] "Say one of my colleagues says something that leads me to think he's insulting me or not showing me respect. Then I feel, *I've been wronged by him*. That's very important in escalating my anger, the feeling he's done me wrong—then I think, *I'm the victim*. So I get angry and want to retaliate and punish him.

"But I've learned from personal experience that 90 percent of the time when I get into negative thinking like this, I'm wrong. So then I start to question my thoughts. Is there some other explanation for what he said? Maybe he doesn't feel well. Maybe I didn't hear him correctly. Or maybe what he says is true."

That resonated with the Dalai Lama, who pointed out, "All these negative emotions are based on misconceptions, on a big gap between appearance and reality. Education means narrowing that gap."

Beck elaborated on how to narrow that gap using cognitive therapy: "So then I explore these possibilities. I might realize he's in some pain himself, which gives me a feeling of compassion. I think, *He couldn't have said that to me unless something was really hurting him*. As soon as I have that understanding, my anger reduces."

That method struck the Dalai Lama as quite similar to "analytical meditation," a practice that uses an investigative inquiry—and, I would add, offers another tool for mind whispering.

Such inquiry into our hidden-mode assumptions might happen naturally when we talk over our worries with someone we trust, who then offers information or a new perspective that lets us see what we thought was worrisome in a less threatening way or that our reaction had more to do with our own exaggerated fears than the actual situation.

In cognitive therapy for panic disorder, such correctives need to be repeated and repeated, because changing mode habits—and their underlying brain circuitry—takes time. For someone with panic disorder, that corrective could come with the simple assurance that, contrary to his knee-jerk assumption, he will not suffocate if he goes through an automobile tunnel. As the client reminds himself of this while connecting with his secure mode, his panic about the tunnel gradually weakens.

Once we surface the assumptions that have kept us hooked into an unwanted mode, there's another helpful approach we can use. Some cognitive therapists have their patients carry small flash cards that state the arguments against their familiar skewed thoughts. The patients write these in reflective moments when they can think clearly and consult them in the heat of the moment when they can't think. A convincing, more realistic belief, says Beck, "forms a solid protective wall against subsequent mode attacks."[2]

A mindful inquiry allows us to be aware of the ways we are thinking while in an unwanted mode and to detect the usually hidden algorithms that lie at its heart. By bringing those beliefs out into the clear light, we have a unique chance to challenge and gradually change the twisted assumptions driving a mode. We can rewrite the narrative.

Perspective

Maya was staying at an inn for several days while on an assignment for work. She loved the place, but one of the women who worked there was getting on her nerves. Maya would be sitting in the lounge, getting herself some tea so she could settle in and work, and the woman would ask, almost suspiciously, "What are you doing?" After a few days of this, it seemed to Maya that the woman had decided to give her a hard time.

Maya tried to avoid the annoying woman. But one day the woman again tried to micromanage Maya and in a rude way that triggered Maya's anger. Maya's first impulse was to retaliate by making an angry remark. But instead she said to herself, *Give yourself a time-out.*

She went back to her room, took a hot bath, and curled up with Facebook. She saw that someone had posted a quote saying, "When you are caught in an angry reaction, you don't see clearly what's going on. You only see your angry feelings."[3]

Sometimes a perspective shift is all it takes to change your reaction.

When Maya read that, she softened and started to consider other possibilities. She realized that the woman had looked very unhappy and that she may not have been directing her bad mood toward Maya in particular. Maybe she was just having a bad day.

Besides, Maya had felt disappointed by a colleague earlier in the morning, and the woman had been a handy target for *her* frustration. Once she saw more clearly what was fueling her frustration, and changed her perspective, Maya relaxed as her mind became more spacious. She let it go.

Maya derailed her anger by applying three methods: a time-out, to cool down; a discerning investigation into her own state; and, perhaps most powerful of all, taking the other woman's perspective.

Being Present with Heart

Sue's problem with Mike is how he treats their daughter, June. As Sue puts it, "They both have big personalities and they clash. Mike just can't see all the wonderful things in June. He only sees the negative. I know it comes from his childhood—Mike's father treated him the same way—but it's still painful to watch this go on in my own family."

Sue feels protective of their daughter and sticks up for her whenever Mike starts an attack, like when he puts down June for texting her friends instead of doing her homework. And Sue judges Mike harshly, which has created an emotional barrier between them.

"It's tense for everyone," Sue says. "Mike just can't control his temper, and I'm constantly blaming him and worried about what's going to happen."

Sue decided to meditate daily. "The mindful overseer helps when I'm sitting. I can hear my mode voices loudly and clearly see how they lead to reactivity. This is more obvious to me when I'm reflecting than in the heat of the moment."

Her daily meditation includes focusing on lovingkindness for Mike and for herself. Since she has started this, she feels more patient with him.

When they talked the problem over, she saw him in tears about his own troubled childhood, and she felt deep love for him at that moment. The image of him admitting his vulnerability helped Sue feel a tender compassion for him rather than flip into negative judgments.

"I was able to focus on his pain," she said. To remember him in pain helped her attune and connect with the love she feels for him.

Now Sue is less reactive. Instead of arguing with Mike in front of their daughter, they find a quiet time to talk things over.

Along with mindfulness and lovingkindness, Sue has found investigative inquiry helpful. This focuses on questions like *Where am I right now?* and *What's keeping me from being present?* This helps us not only recognize which mode we are in at the moment but also to take the next step and explore what remedy might be needed.

"I realize how I'm prone to the perfectionist mode and I'm more able to recognize when I'm caught in it," she tells me. "I've been thinking about Mike and how to respond to him, and it's helped to see myself as less judgmental and more loving and kind.

"I'm seeing my modes more accurately," Sue says. "I can see how I am in my adult, secure mode. I'm different at work and home—secure at work and judgmental at home. I'm more aware of when I'm starting to slip into the perfectionist mode. As I'm losing it, I catch myself quicker and I remember: just be present with heart before my judgment takes over."

We can't change the past, but we can change how we react to the present. Instead of trying to change the people in our lives, we can transform the patterns between us. Sue was able to see her husband's needs as well as her own and take steps to unhook from relentless reactive patterns.

Our habitual negative modes can be ingrained and very hard to shake, let alone see. Those with the most emotionally intense reactions call for a more lasting transformation, one that modifies the underlying structure of the mode.

Our extreme modes are parts of ourselves that we get stuck in. Mode work helps make whole these fragmented parts of ourselves by integrating their better aspects so we can move smoothly from one to another.

When it comes to dealing with modes that come back to haunt us over and over, instead of just trying to shift out to a more desirable mode we can choose, as Sue did, to engage it: to explore, understand, and heal it from within.

Along the way a series of obstacles may arise: distractions, strong reactions, fears, anxieties, doubts, avoidance, or feeling a lack of skill or knowing what to do. But as we gain glimpses of how to pull ourselves past those obstacles, real change takes root.

In this process, connecting with our secure mode can give us firm footing in an investigative awareness—so we are less consumed by our distorted assumptions and overreactions—which makes us more able to experience upsetting feelings without being overwhelmed by them. Sue told me at one point, "Knowing there's a secure mode to connect with has been so reassuring."

What Is Behind Our Modes?

Alcoholism, drug abuse, and other addictions— extremes of the attachment mode—can be seen as attempts to sedate distressing feelings. A psychotherapist in Los Angeles told me that she uses methods from my work with her clients who are recovering from addictive disorders.

The methods help them become mindful of their triggers. Then investigating what their real underlying emotional needs are gives them a choice point they had not known they had before: a greater ability to tolerate their cravings without acting on them. Instead of reflexively reaching for a drink or a drug, they could stop, look inward, and find a remedy at the emotional level—whether talking to a friend, journaling, exploring how past needs were playing out in the present, or connecting with a greater awareness that felt nurturing.

There are two basic ways to respond to our suffering, says the Dalai Lama. "One is to ignore it and the other is to look right into it and penetrate it with awareness."

Sustaining an inquiring presence to an extreme mode like addiction takes a no-bullshit clarity. The same goes for the other modes. The anxious mode might, for instance, be a way to cope with fears of abandonment. Can we challenge the automatic rumination on fears about our personal connections and find healthier ways to respond? What is this pattern telling us we need? Are there better choices within our grasp?

As we connect with what underlies a mode, we are likely to find emotions that may need to be acknowledged, understood, and digested. Then these feelings can be integrated instead of avoided, clung to, or solidified.

Our emotions can harbor wisdom, if we can yield to it. If we reject the emotion, we risk cutting ourselves off from what we might have been able to learn. But sustaining attention with a mindful inquiry allows us to be more present with feelings without our minds wandering off as we look for an exit.

Investigative awareness goes beneath a mode's story line. Allowing an empathic investigation can sometimes penetrate to the energy driving the feelings, and so new meanings can be revealed. But given our human tendency to avoid strong feelings, this takes trust in the possibility that exploring emotional responses can help. We need to first recognize we can be safe facing long-suppressed feelings. As that investigation changes our understanding, we are more able to engage emotions.

Sustaining awareness even amidst a storm of rocky feelings is a crucial part of mode work. If we lose our connection to inquiring presence—say, by getting completely lost in an irrelevant train of thought—the knowing quality of awareness wanes.

The investigative beam of discernment allows us to experience an emotion without being overwhelmed by it, so we can inquire into and integrate new information. When you investigate, you let more of the truth of emotions into your awareness—but do it in a way you can manage. With more intense feelings this can be done by shifting back and forth from turbulent feelings to a calming practice.

A student who used to live in Hawaii put it this way: "Mindfulness of modes feels like surfing a powerful wave—it can help you stay more centered and balanced while riding the energy of an emotion."

This is not to make it all sound easy—there can be compelling reasons we reel with upset or writhe with pain. Life can be difficult, in particular ways for each of us. There is not one set prescription for everyone. Each of our predicaments is unique and requires its own remedies, in addition to the general ones, like care and emotional support. We don't always know how to find our way through or know what we're supposed to be learning from life's challenges.

At the core of each of our distressing modes lie conflicts and deep needs; unsatisfactory solutions to these needs are what make a mode maladaptive. A longstanding inertia and complacency holds these patterns in place. With an attuned discernment we can ask, *What am I really needing? Can I find a better solution?*

What allows the lotus to bloom through the mud?

Boundaries

A client felt for many years that her mother successfully manipulated her by either being demanding, helpless, angry, or guilt-provoking. This triggered self-sacrifice in my client, who would give in to an unreasonable demand or put up with some inconsiderate expectation.

One day it was so bad that my client found herself pacing back and forth anxiously. Then, she told me, it suddenly dawned on her, "I feel like a prey animal and my mother is the predator!"

With this insight she calmed down immediately, and her anxious fears gave way to calmness, where she could begin to think more clearly. She saw what was needed: boundaries. Her mother had to learn how her controlling behavior was no longer going to make her daughter self-sacrificing prey, that there were consequences to her predatory ways.

From then on, my client was mindful of times when her mother was starting to get manipulative again and she was steadfast in maintaining her boundaries. This change in their predator–prey standoff improved their relationship greatly.

As we become aware of subtle predatory behavior in our relationships, this very awareness creates a choice that opens up new options. We can respect the person while not accepting how they treat us. Otherwise, the predator-like mode perpetuates itself, as habits do, and can erode our relationships.

Once we have increasing glimpses of what a more genuinely empathic, caring connection feels like, that becomes extremely motivating. Our reference points shift and we can become disillusioned with such dysfunctional modes of relating, preferring kindness, trust, depth, and intimacy.

This begins to change your connections. For example, consider communicating skillfully with people who have a subjugation-triggered mode. Such people are hypersensitive to feeling controlled, so relating to them by considering their wishes and giving them choices prevents priming their mode.

But then, why shouldn't we treat everyone this way? Nobody likes feeling micromanaged or controlled, let alone manipulated. It seems far better to make joining up our default mode, no matter whom we're with and what we're doing.

In the Danger Zone

I don't want to give the impression that we can always free ourselves from the hold of a toxic mode entirely on our own. Sometimes it can help enormously to work with a trained professional.

For those who find themselves trapped in modes so extreme they have become debilitating, I recommend Jeffrey Young's mode-focused therapeutic intervention, which he calls schema therapy.[4] Other approaches to psychotherapy are, of course, helpful too—but his is the method I'm most familiar with that explicitly targets mode problems.

Jeffrey Young's insights bring more understanding to modes while changing deep habits. His approach brings to the surface the underlying insecurities that shaped a mode when it developed; for instance, the patterns of entitlement (an aspect of the predator-like mode) are seen to be compensations for vulnerabilities like feeling unlovable or emotionally deprived. Young distinguishes between pure narcissism, seen in people who as children had no limits set for them, and fragile narcissism: the vulnerabilities are seen in the fragile variety.

At the same time, Young emphasizes, people with this mode need to learn the negative consequences of their self-centered behavior and to have empathy for the pain it causes others. While therapy can help them heal the underlying emotional wounds, at the behavioral level they need to learn to respect boundaries that limit what they can do in order to protect other people. This multi-layered approach offers a model for working with other extreme modes in a way that heals the underlying emotional vulnerabilities that have led to the mode in the first place, as well as changing mode habits.

Schema therapy sees the healthy adult dimension of the secure mode as having a capacity for some inner re-parenting.[5] In this mode we can soothe the sometimes-desperate needs of our less healthy modes, such as the anxious mode's need for security. We have the courage to face unsettling feelings that the avoidant mode dreads or challenge the maladaptive habits of our less desirable modes.

During psychotherapy the healthy adult becomes stronger as the therapist models being nurturing, validating, affirming, and supportive. This encourages clients to rely on this mode internally. As we do so, we are integrating fragmented parts of ourselves.

Roger went into schema therapy. He was a seasoned mindfulness meditator, but his practice hadn't seemed to help with the modes that were destructive to his marriage. He and his wife, Suzanna, loved each other, but over and over they'd get into one of their toxic arguments, which always left them filled with tension for days.

His mode homework included spotting his bad modes whenever they were triggered and applying a remedy. As he started applying mode mindfulness he realized that he and Suzanna really only ever had two or three basic fights, each a variation of the same bout of mutual mode-triggering.

Once, for instance, they were getting ready to go on a trip and, as Suzanna was packing her carry-on luggage, he said to her, "That's too much. You better leave a lot of that behind."

At that, Suzanna snapped back with an angry tone, "Don't micromanage me."

And Roger, in turn, simply froze. He felt intense anxiety. And the rest of the evening Roger and Suzanna were out of sorts.

The next morning Roger realized that when he had made that remark he had been in his duty-bound mode, which makes him critical and controlling. And he recognized the resulting sullenness between them as an all-too-familiar impasse, a silent mode war.

Roger saw that his remark about packing, and a bit of micromanaging, was impelled by his perfectionistic, duty-bound mode, and that Susan had responded from her rebellious prey-like mode. That, in turn, brought out Roger's prey-surrender mode, one in which he just gives in, making him so frozen he does not know what to do next.

Later that day he had a session with his therapist, who was pleased that he had recognized the modes at play. "It would be better," Roger's therapist advised, "if you shared your own feelings. Speak from your heart, admitting your fears and concerns."

When Roger then told Susan about his insights into their tense exchange, he learned something else about their mode dynamic. "When you get into that duty-bound mode," Susan told him, "you get so controlling it drives me crazy. I was reacting to feeling controlled and uncared for. I had been feeling so open and loving with you before, but when you said that, it shut me right down. I felt you cared more about the bags than about me."

At their next meeting, his therapist helped Roger think about an alternate way to express himself. She advised him, "You could say, in a caring voice, 'I'm feeling a bit concerned about you. I'm worried that your most valuable things—the ones you feel you need to carry on the plane—will make your hand luggage too heavy and that you'll have to check them. And I know you've lost checked bags before.'"

"Then," his therapist added, "you can just see what your wife has to say, and let her find her own solution. That way she will feel cared about, that you are not trying to control her life."

It's not so easy to come up with these mode antidotes in the heat of an argument. Marital researchers recommend that arguing couples take a time-out for at least twenty minutes to calm down before they come back together to try to patch things up.[6] But the subsequent patch-up will be far more likely to succeed if during those minutes of de-escalation each partner (or at least one) recognizes the mode that has consumed each of them and takes remedial steps, ideally accessing the secure mode even while working out their differences.

Getting into the habit of recognizing their favored modes at work during their arguments can help a couple notice a trigger in the first place—and therefore not have the resulting fight. Roger's mode therapist gave him some step-by-step advice to use when he sees he's about to enter a mode-triggering danger zone:

- Learn your mode signals and tune in to what is driving you. Notice what mode you are in at the moment—pay attention to thoughts, emotions, and sensations. If you sense a bad mode, work on confronting the mode itself.

- Think through what to do before you react.

- Reflect on how the other person might be feeling and thinking.

- Think of alternatives. (For instance, Roger could have told himself, *There might be a problem with how Susan is packing for this trip, but maybe I'll discuss it with her when she's not so pressured.*)

- Remember the remedy, and then act. (For Roger that would be first expressing his concern for Susan, then giving her the facts, but without being controlling; let her come to her own resolution.)

The study of epigenetics has revealed, in essence, that we may possess the genes for a given disease, but if, for example, genes for a disease like diabetes never get triggered by eating lots of carbs and sugars, and we stay fit, we're far less likely to develop the symptoms. In Western preventive approaches this principle can get quite targeted, starting with analyzing the earliest causes, such as chronic inflammation, a pathway to many diseases.

In some Eastern healing approaches, treatment starts with diagnosing the underlying causes in terms of modes, such as attachment, aversion, or ignorance, which are seen as leading to emotional conflicts as well as physical ailments.

This contrasts with conventional medicine's focus on symptoms rather than their early precursors. But why not address the initial causes and avoid triggering the pathological processes in the first place?

Likewise, mode work aims to get at the root causes of our most extreme modes rather than just manage their symptoms or simply shift us out of them. By targeting a mode's underlying causes we can prevent the emotional tendencies that otherwise might build up to full expression in disturbing modes. Think of it as epigenetics for the psyche.

Voices of the Mode

Liza, who was prone to prey-like self-sacrificing tendencies, went through months of feeling deep resentment toward anyone in her life who she felt was self-absorbed and ungiving. She found herself hypersensitive, overreacting to the smallest hint that people were being uncaring or not reciprocating for all the times she'd gone out of her way for them.

Liza felt angry and frustrated. Her distorted mode thoughts led her to turn even innocent events into flare-ups, amplifying her fears of "being used" yet again, even when there was only the slimmest shred of evidence.

She had reached her saturation point. She was tired of the situations that triggered her, of whatever reinforced her patterns, and of the mode itself. Such exaggeration of your mode reactions can be a signal that you are reaching a tipping point, that you have become so disillusioned with the mode, you are highly motivated to change the patterns.

A period of heightened mode intensity—with the voices of the mode repeating more often and ever louder—can be a useful stage in working through and healing a negative mode.

It can be a good sign, but only if we see clearly what's going on. Liza started to recognize her mode's voice in strong, repetitive thoughts like, *After all I've done for him he's so ungrateful,* and, *She doesn't really care about me.*

These thoughts flow from the distorted beliefs at the heart of the mode. If we're unaware, they simply become self-confirming: *See, it happened again. That proves people just want to take advantage of you.* But with awareness, we can begin to recognize the thoughts as voices of the mode, not reality. We can see how we've believed the thoughts and how the feelings they evoke have fueled our responses for so long.

If Liza hadn't recognized these inner voices as characteristic of the mode, she would have gone on believing them. As we become familiar with the voices of our mode and can name the thoughts as such, they begin to lose their power over us. This awareness shifts our relationship to the mode thoughts; instead of instantly believing them, we can feel disenchantment with the spell they have held over us.

For Liza this recognition let her see the sequence, repeated over and over: her own actions fueled the core beliefs of one quality of the prey-like mode, self-sacrifice—triggering a toxic mix of resentment, indignation, and hurt—along with story lines like, *People just want something from you. They don't really care about you.*

Liza was flooded with thoughts like this when she realized they were the voice of the self-sacrifice mode. And then she had the insight that they were escalating because she was ready to release the pattern. Modes want to survive. Like other deep emotional patterns, they can get stronger when we start to focus on them in order to heal.

She could see how the distorted thoughts and over-reactive feelings of her mode were driving her emotional habits—in her case, either rejecting people for their imagined slights or engaging in overly solicitous caregiving. Such an insight offers a choice not to let the relentless pattern continue.

With modes that carry a strong emotional charge, that choice takes a special kind of work. If we are to change the mode patterns, we need to engage with the underlying emotional dynamics that hold it in place. These patterns have been basic to who we think we are. As we let them go, we may find we have to grieve for the lost parts of ourselves we have been defined by.

Liza said at one point, "I stopped taking care of people all the time and I wasn't sure if they would like me anymore. Some didn't like it, and I saw that I had to be willing to let go of relationships that were based on my being constantly the caretaker in exchange for being liked. But the true friendships last, and I feel so much freer not being driven by this pattern."

The Overseer and Habit Change

In mindful habit change, the overseer

- checks in;
- knows how modes think, feel, act, and interact;
- takes the mode temperature;
- sees what's needed;
- knows the various interventions, and which might help; and
- chooses accordingly.

For example, a web designer had a huge deadline looming, and a long-planned family gathering was coming up during which her relatives would be staying with her. The designer felt she really should work right through the weekend of the gathering, but she didn't want her family to feel hurt. She was self-sacrificing around her family without giving it a thought, putting aside her own needs—even a major deadline—to be there for others. But she was already feeling overwhelmed even before her family had arrived.

She was exhausted and had been ignoring the signs of her body's need for a restorative break. She was so wound up she tossed and turned much of each night, then would force herself to keep going during the day despite her low energy.

Then, when her family was just about to arrive, she remembered to tune in to her mindful overseer, and ask, *What's happening?* That launched an internal dialogue between her overseer and her mode. She thought about her houseguests and their needs and how usually she would feel fine accommodating them all. But now that she was exhausted, with her diminished energy, it would be just too much.

She acknowledged her self-sacrificing strain of the prey-like mode, recognizing those familiar patterns of thought starting to assert themselves again. She sensed that this time it felt different: she was seeing things through the lens of real exhaustion. She noticed that she was feeling a bit resentful of the coming guests, and she assumed they would feel entitled to her hospitality.

At that point her mindful overseer interjected: *What are you needing?*

The immediate response: *Rest, space, and reciprocity.*

She had a memory of once when a former houseguest had been going out and had asked, "Do you need anything?" and how amazing it had felt to have her needs considered and how rarely she felt that way. She recalled, too, that her self-sacrificing core belief leads her to assume that people are more inclined to receive than to give back, which added to her feeling of being overwhelmed. *This mode pattern is tiring,* she thought.

Then the mindful overseer reminded her that the self-sacrificing mode stems from a learning background. The people around her had lots of unmet needs, and she had learned to cater to them for fear of losing those connections. The mode led her to overextend herself in order to satisfy her own relentless emotional needs (to be loved, accepted, not abandoned), which was a bottomless pit. It always brought disappointment, because she didn't feel appreciated—everything she did was just expected or even demanded of her. She felt like some useful object.

Her inner overseer reminded her of a mode remedy: boundaries. When people in this mode start looking for space or relief from the pressures, they feel they can't continue without boundaries—not transparent ones but wetsuit-like, waterproof boundaries that nothing can get through.

The mindful overseer suggested, *This mode needs to redefine relationships in a healthier way. There's something called "clear communication" that can be helpful in times like this.*

How does that work? her mode responded.

You can be clear and nonreactive when you communicate, the overseer reminded her.

Great idea! replied the mode. *I feel a weight being lifted already. I'll try it.*

At that moment this inner reverie was broken by the arrival of those houseguests. The web designer decided just to be honest with them. So when her sister-in-law asked, "How are you doing?" and gave her a big hug, the designer said, in all honesty, "I'm feeling overwhelmed. I have a huge deadline I have to meet, and yet I really want to be with you all."

They talked it over and a little while later her sister-in-law said to her, "You know what I felt when you told me you were overwhelmed?"

The designer thought, *Oh, no. I hurt her feelings. And she's traveled so far. I shouldn't have said anything.* The script was straight out of her self-sacrifice pattern.

But her sister-in-law said, "Relieved!"

"Relieved?" the woman replied.

"So often when we come to visit I know you put things aside to take care of all of us, and I always feel we're being an imposition. You're always so nice about it, but I can sense your internal pressures. When you told me you felt overwhelmed, I found I could think of ways to be more helpful to you and less of a burden. You appear perfect, but you're just human."

This was an eye-opener, and a mode remedy, for the woman. She saw her self-sacrifice patterns—appearing not to have any needs of her own—as creating a distance in her relationships by not letting people know her needs and be there for her. This was a first, small step toward mindfully transforming her mode habit, and she felt more genuinely connected to her sister-in-law.

Mode Correctives

Leslie didn't hear back from a friend she had texted, one who usually responded right away. That triggered an anxious mode and the conviction, *Maybe she's mad at me.* Leslie felt hurt. She assumed her friend was being cool and distant and wondered if she had done something wrong. She began to spin in anxious ruminations, casting about for ways to assure herself that the connection between them was secure.

Then Leslie remembered other times she had been like this. She paused and said to herself, *This is my anxious, clinging mode at work.* Then she recalled that her friend's son was ill. And with that, the more objective insight, *She's not ignoring me. She's just preoccupied.*

Leslie stopped herself from acting on her anxious fears of disconnection, which had led her again and again to reach out for reassurance. That would just reinforce the habits of her anxious mode. Instead, she decided to give it a rest for a while.

A negative mode's distinct core beliefs give rise to predictable automatic thoughts, handicapping inner voices that express distorted assumptions like *I'm not safe in this situation* or *I'd better distance myself before I get hurt.*

For this woman's anxious mode, the reparative lesson was to counter the tendency to exaggerate fears and not to be controlled by the assumption that distance in a relationship means the end of it.

As we're working with our more distressing emotional patterns it helps greatly if we can access our secure mode, with its empathic awareness and clear discernment, to give us a place to stand while we dismantle the negative mode's constructs. A mindful pause can sometimes help us access our secure base long enough to look into our thoughts and find a remedy.

Such correctives slowly change how we appraise mode triggers. Our usual knee-jerk responses are undermined as we see things more clearly rather than through the mode's distorted lens. As we weaken the habitual mode response, we can neutralize it with its antidote.

With repeated corrective experiences, we can more readily stabilize enough in a secure mode that we can face what had been mode triggers without feeling destabilized. This marks a lasting alteration of the dysfunctional mode's underlying beliefs, a change in our learned responses, and the defusing of its triggers.

As we mindfully reappraise events, there are changes in our brain akin to hitting the REFRESH button on a laptop. Of course deep emotional habits don't change overnight. Working with modes can sometimes seem like taking one step forward and one step back—like rap dancers moving in place.

Each mode has its own set of specific antidotes, the correctives that help on the path to integration and transformation. The antidotes for our negative modes are very different from what the mode "wants."

For instance, the avoidant mode seeks to distance us from strong emotions, fearing we will be overwhelmed. But to heal the mode we need to learn to open ourselves to those emotions and to feel comfortable doing so. We need to learn and master new habitual responses that replace our old, dysfunctional mode habits.

For the modes of attachment and aversion, the question is: Can you experience pleasure without attachment or discomfort without aversion? The troublemaker is clinging to wanting or not wanting.

Again, there is no universal list of antidotes for each mode. For one, we all experience them in unique ways. Still, there are rough general guidelines that can be helpful to keep in mind, especially when we are trying in the heat of the moment to recall what a given mode might need to learn:

ATTACHED: to apply restraint to impulse, to be self-contained and content; to be generous and disenchanted.

AVERSIVE: to be patient and accepting, and to empathize and understand others.

BEWILDERED: to bring clarity, investigative inquiry, and a discerning attention to situations.

ANXIOUS: to challenge exaggerated fears—particularly that a distance means you are losing the connection to a person—and to be more self-contained in relationships.

AVOIDANT: to feel safe confronting intense emotions, and to become more comfortable with intimacy and closeness.

PREDATOR-LIKE: to empathize and consider other people's needs and wants rather than imposing an agenda, to be open to honest feedback, and to take responsibility rather than blaming others.

PREY-LIKE: to be assertive in expressing your needs and rights strongly and clearly, and to connect with your own preferences.

PERFECTIONIST: to relax and be more accepting of yourself and others, to be more spontaneous, and to balance productivity with taking care of yourself.

Knowing what a mode needs in order to heal gives us some guidance in making wise choices, a key step in mindful habit change.

CHAPTER 12

Priming Our Secure Base

When I was seventeen, a couple hired me as a mother's helper to take care of their infant. One night the parents were out, and the baby and I both fell asleep early.

A few hours later the infant's hysterical crying woke me. When I went into the baby's room and lifted her into my arms, I could feel her poor tense body shaking from crying so much. Holding her close, I felt such warm empathy for this baby I loved, and I tried to comfort her with my voice. Suddenly, I felt a strong wave of tenderhearted compassion, almost like a surge of energy that seemed to flow out of my heart into her body. As soon as this happened, the baby melted in my arms. Her tiny body became heavy and limp, and she fell fast asleep.

Such moments are routine for parents caring for an infant, but as I was still a teenager, I had learned something new about compassion. That was the first time I had experienced so vividly how expressing a sincere, tenderhearted love just might help someone find her secure mode.

The very idea of the secure base comes from the model of a caring parent attuning to a child's needs and making that child feel understood, loved, supported, and safe in the world. The people who love and care about us can prime this mode.

If we find partners or friends who are sensitive, responsive, and caring, having repeated secure-base experiences with them can be reparative, making us more able to be there for the other people in our lives. Our views of human nature take on a more positive outlook as our negative modes go limp.

It can make a huge difference to have someone who is interested in and cares about us when our primal needs are unmet. But if there is no such other person, it's not too late to connect with those qualities within our own minds and hearts.

There are two doorways to the secure mode, one inner and the other outer. While we can turn to loving people to prime this mode, we can also look inward. There are many ways we can build the foundations of a secure mode on our own and become that source of nurturance for ourselves.

When a gardener tends to plantings, a host of conditions must be in place for the plants to flourish: among them, tilling and fertilizing the soil, creating beds, seeding, watering, weeding, and protecting the seedlings. The more such tender loving care, the more the plants will bloom. The life-giving force you dedicate to caring for the plants in turn yields their growth.

Likewise, we can nurture the qualities of our secure mode by creating the conditions that let this inner safe haven flourish. Joining up in connectedness with nourishing people can be one, and so can joining up within. There are different levels of joining up. Our distorted modes are patterns that disconnect; our secure mode connects.

Acts of kindness, clear communication, caring concern, and empathic attunement all prime our own secure base. So does nurturing our positive qualities, finding meaning in our lives, seeing things with an accurate discernment (rather than through a distorted lens), and creating safe inner harbors. The more we use these internal paths, the greater our confidence grows in our inner resources.

Mode work itself frees our minds and hearts and lets the life-enhancing secure mode emerge as our default stance, the place inside that we return to over and over.

Beyond repairing our individual modes, this work is also about holding the view of interconnection; our confused modes obscure that view. As we transform our perspectives, we can more often live in ways that express that view.

Safe Haven

That epiphany with the crying baby came during the same period my first boyfriend gave me a copy of Hermann Hesse's *Siddhartha*—the novel based on a spiritual seeker who encounters the Buddha—and then went off to college and broke up with me. At the time, I couldn't imagine being without him; this was my first heartbreak, and it primed all my abandonment fears.

I dove into that novel as a profound refuge. I found solace in passages like the one where the seeker Siddhartha comes upon a river and a voice within instructs him to sit there and learn from it. He saw that "the water continually flowed and flowed and yet it was always there, it was always the same yet every moment it was different."[1]

Such insights helped me reframe my heartache within a larger dimension, one that affirmed the permanence of change, the nature of suffering, and the attachment that underlies it, which in turn helped me get over the loss of this important relationship.

Even years later, when I once again spent time with that first boyfriend, we considered getting back together.[2] As I look back to that time when I was seventeen and recall the pain of that separation, I now see it as freeing myself from my attachment, which led to other worlds unfolding for me. If I hadn't been willing to let go, hadn't been open to change, I would have missed those opportunities.

Those weeks and months of emotional hardship fostered a new direction in my life; the adversity transformed into opportunity. I began to meditate for the first time. Having the guidance of these teachings, while going through this painful time, made me wholeheartedly plunge into meditation practice. I found I could connect with the inner refuge of a secure mode.

That led me to intensive meditation retreats, travel to India, my connection with wonderful meditation masters, and my eventual enrollment in a graduate program integrating Eastern and Western psychologies, which in turn led me to the work I do now. I sometimes feel like an inner tour guide, encouraging others to connect with the adventure within and free their minds and hearts.

Practice has been such an illuminating path to emotional freedom, awakening deep insights into my life, that I have particularly felt the need to share the benefits of meditation. These include redefining our limited sense of ourselves and of others, and embracing an expanded, more spacious view of our world. A compelling fruit of meditation practice for most of us may be in finding a path to the inner refuge of the secure mode.

The secure-mode benefits of mind training include finding rich inner resources, such as feeling replete and self-contained or being more accepting of things we cannot change and feeling less need to control what we cannot. We get a more balanced perspective, one that gives us a larger view of events; we can see the sky behind the clouds—or at least remember it's there.

These are all qualities we might hope our childhood caretakers had brought to us early in life. But whether or not they brought these secure-base qualities to their caretaking, these are still qualities we can nurture and develop within ourselves. In this sense, training in meditation is a form of inner re-parenting.

Among the many other qualities of the secure mode that meditation enhances, a few stand out. For one, we become less dependent on externals to determine our inner state as we anchor our attention in a larger awareness, one not defined by our outer condition. Our sense of calmness and an inner security grows stronger as we turn toward a nurturing awareness within, instead of depending on other people for this.

Calmness can grow into equanimity during life's turbulence, which gives us a place to stand within, a nonreactive awareness, and a balanced perspective. What might otherwise have been a negative mode trigger becomes just a neutral part of life's passing show. Then there are a range of positive feelings, such as generosity, resilience, and playfulness.

Enhancing our focus and clarity lets us access more of our minds' true potential. We have more room to see clearly, whether our own emotional issues or deeper insights, like natural principles that govern our experiences. Sensing more strongly the impermanent nature of things helped me reframe breaking up with my boyfriend as a part of life rather than some overwhelming tragedy. This freed my spirit and opened up other possibilities for finding my own direction.

Along with a calm clarity come an understanding and a sense of compassion for the suffering caused by our own—and others'—distorted mode perceptions. We can see and acknowledge the poignancy of our shared human condition.

We gain a greater capacity for inner attunement and a growing confidence that we can take charge of our internal worlds—that connecting with our true nature in order to gradually make our minds freer is a real possibility. And that same capacity for attunement translates into more genuine connections.

The Power of Love

A few days after her terrible accident, Robin woke up in the hospital's intensive care unit and realized her mother Diane was at her bedside. "It really helped that my mom was there; just having her nearby was very touching. I thought she was only going to be there for a few days, and she ended up staying the whole five weeks with me, during my time in rehab. She was always being supportive, by watching me do my physical therapy and giving me positive feedback and being my advocate, handling all the paperwork, insurance—things I couldn't do myself."

That intimate closeness with her mother no doubt strengthened Robin's sense of the secure mode. The people in our lives who we love—our families, our close friends, and even our pets—all shift us toward the secure mode.

The mode-shifting power of the mere presence of a loved one was discovered in an experiment in the brain lab of Richard Davidson at the University of Wisconsin. Women had their brains imaged while being told they were about to get a mildly painful shock. Their apprehension was evident in the heightened activity in their amygdalae and other parts of the brain's circuitry responsive to danger and alarm. But if a woman's husband held her hand during the conversation, the woman's amygdala quieted completely.[3]

A client once told me, "Every morning when we first see each other after waking up, my husband and I give each other a warm hug. It gives me a sweet sense of connection." She added, "But he puts it a little differently: he says our little ritual floods him with oxytocin, setting his secure base for the day."

That hug seems the human equivalent of what researchers have found when female rodents lick and groom their offspring, which appears to set the offspring's genes toward the secure mode. There are many ways we get this licking-and-grooming equivalent in our lives, such as cuddling with a loved one or connecting with our children or anyone else who opens our hearts. The nurturance of a close friend or confidant can give us the reassuring feeling of being accepted, supported, and loved. When we feel destabilized, just talking with someone who is caring and feeling his or her support can enhance our secure mode.

Feeling that others have an empathic attunement, with the tender support of a caring heart, can be a powerful secure-mode prime. When we see someone in need of such connection, we might do whatever we can to help her connect to her inner strengths; we become something like each other's immune system of the heart.

Modes, like moods, are contagious. Someone who is connected to her own secure mode can be a soothing influence on us, simply through her mere presence. To send positivity takes some stability. Otherwise we are more vulnerable to receiving whatever may be emanating from those around us.

As we become more familiar with our own modes, we also naturally develop a greater awareness of modes in other people. Seeing modes more clearly in someone else gives us an opportunity to do for that person what we are learning to do to help ourselves.

Say that your friend is under the spell of the anxious mode, hooked into being upset and overreacting. You can help him simply with your mindful presence, offering him the safe container of the secure mode by paying full attention with heartfelt empathy.

It's not advisable to tell him "You're in an anxious mode" and try to talk him out of it; while he's caught in the mode, little will get through. That kind of response to someone's fretting reflects cognitive empathy alone, devoid of emotional attunement.

Instead you can yield to a sense of calm spaciousness. Just as you would let go of your own mental hooks, you can do so now by not becoming reactive in response to him but intentionally just letting things be. A kind and caring warmth can help him feel the safe haven you are offering, if only at the subliminal level of neural resonance.

No matter what you say or do for him as the encounter continues, the atmosphere of that healthy mode will have an effect. This invites him to find that same mode within himself. Connecting and sustaining an inner secure mode, so that it becomes a dependable reference point, helps relationships of all kinds, including that with our own minds.

A friend said, "Yesterday, as I stood glumly in the checkout line at the market, miserable from a cold, the cashier looked at me and asked with all sincerity, interest, and kindness, 'How are you doing today?' It completely changed my day, which in turn changed how I related to my family when I got home."

The people I've been fortunate to meet who have come to my workshops over the years have been an inspiration to me—in their honesty, sincerity, and wholehearted openness. They start as strangers, but surprisingly soon a sincere trust reveals itself and we start to feel like a bonded community.

This connection has happened over and over with people who are sharing stories of their losses and lives, learning from each other, sharing the pain of another's heart and the inspiration from each other's insights. It's become so clear how we are from the same human family, sharing the same essence.

And when people are willing to be themselves, engage in honest introspection, and share their human vulnerabilities as well as their triumphs in facing adversity, we create a comforting secure mode through our common humanity. Everywhere everyone gravitates to a shared secure base.

Look for the Helpers

That surgeon who endured a trial for malpractice countered the inevitable anxiety by creating a secure base: getting under the covers with her husband and having a pizza. Her instinct was right on target for priming a secure mode: being with a loved one in a safe and cozy place and eating comfort food (of course, these days, it should be the low-fat kind!). There's another, internal doorway that results in the same effect. Bringing to mind people and situations that we find comforting gives us a reassuring sense of connection.

There are three varieties of people in our lives who prime a secure mode. Think of the answers to these queries:

- Who do you like to spend time with?
- Who do you turn to for comfort when you're upset?
- Who do you feel you can always count on?

These are the people you depend on, enjoy being with, and know will be there for you.[4] But you don't need to be in their physical presence to benefit; they comprise your inner secure base. Just bringing these figures to mind can be soothing in times of distress, as does recalling a time when one of these people helped you in some way. The inner figure can have the same reassuring, calming effect as an actual person does.

A YouTube video has this secure-base priming effect. A boy falls off his skateboard and, as he is lying on the ground, a man walking by stops and helps him up. A few moments later the boy sees an elderly woman carrying bags of groceries and struggling to cross the street. The boy helps her with her packages so she can cross safely.

This passing forward of good deeds goes on in a human chain of thoughtfulness. Watch a person doing something kind and generous in turn—the opportunities are always there if we're alert to them. This spreading goodness creates what amounts to a secure base community.

For decades Fred Rogers hosted what was at the time every toddler's favorite TV show, *Mister Rogers' Neighborhood*. He embodied a secure base, a warm and caring grownup readily available to millions of kids.

"When I was a boy and would see scary things in the news," Mr. Rogers once said, "my mother told me, 'Look for the helpers. You will always find people who are helping.'" That reassurance redirects attention away from anxiety-provoking triggers and toward security primes—sage advice whatever the specifics of the "scary things."

Phillip Shaver and his colleague, Mario Mikulincer, have studied secure-base priming in its many forms. In one of their studies, people listened to a story about a cozy, secure situation; in another, they visualized that warm, connected situation. Even subliminal methods—superquick flashes on a screen of security words, like "love"—aroused the secure mode. But the more actively involved people were in the method, the stronger the effect.[5]

The powerful message of secure-base priming is that we can cultivate this inner safe haven regardless of whether we have people in our lives available to help us enter this mode.

Think of someone's desk at work and the pictures you recall seeing there. Invariably the snapshots depict the people, pets, or places that person loves the most. They are not just sentimental nostalgia, but something more: secure-mode primes. Looking at those pictures during a stressful day offers an inner oasis of peace, if only for a moment or two, which can help us stay calm and confident, and connected even at a distance.[6]

Much the same effect can also occur when you see or bring to mind images that have similar associations, such as a mother cradling a baby, an affectionate couple, or words like "hug," "love," or "closeness." Seeing or thinking of these primes have soothing impacts similar to actual interactions.

A woman realized how she had been looking for a secure connection from people who were not reliable. That realization helped her find ways to generate an inner security that does not depend on someone else. An inner connection can be nurturing, independent of outer relationships.

Creating an inner secure base on your own is like the magic of connecting with a loving grandparent, caring teacher, or nurturing parent. Your inner secure base doesn't have to rely on a real person; it can be a place where you love being, a spiritual figure, an inspiring leader, or even happy memories.

Energy methods like t'ai chi or qigong can offer another doorway to secure mode. Master Yang Yang, a qigong teacher in Manhattan, teaches a stance that can anchor people in a balanced alignment with strength, a physical analog of a secure base. The feet are placed in a V shape, with one about a foot in front of the other. The body's weight is balanced on this stable foundation.

To practice integrating this in daily life, the master plays rhythmic soft-rock music and has his students dance freely around the room. When the music suddenly stops, the students take this balanced stance. The idea is to find our inner strength and balance in the midst of our lives.

Another source of secure mode can be found in nurturing surroundings. A family with three young kids was living in an old house that had small rooms, particularly the kitchen. Since the kitchen is the heart of any home, they would all end up hanging out in that small space, making the best of it, but inevitably with some friction and sparks.

Eventually they were able to remodel. They did much of the work themselves but made trades with skilled craftspeople for parts of it. They expanded the kitchen by breaking down some walls, and they put in new countertops and cabinets. It had a transformative effect on their minds. They had a cozier, but roomier, nest to come home to and immediately started getting along better. They felt more creative too, cooking new dishes for each other. The space itself seemed to nurture a secure mode.

Nature offers a primal path toward the secure mode. When people are stressed from constantly focusing on work goals or intensive studying, for instance, going out into nature—even on a walk in a park or to an arboretum—shifts the brain into a more open, relaxed state. A walk down a city street, where we have to watch out for traffic and dodge other pedestrians, just doesn't have the same restorative effect.[7]

A business consultant said that for years he had the thankless task of having to explain complex systems to employees who didn't understand them, or didn't want to. He would get frustrated and even lose his temper. So he went to a therapist to get help in handling his anger. At one point his therapist advised him to go somewhere on vacation where he could leave behind all his work problems. He and his wife found a house to rent on the highest ridge of Tortola in the British Virgin Islands. For two weeks they spent hours contentedly enjoying a breathtaking view of billowing clouds, distant islands, and the open sea. He felt a deep peacefulness.

When he went back to work he found he could evoke that same sense of inner tranquility just by imagining himself there, in what he called his Inner Tortola.

When the Mind Expands

Jan and her husband were on holiday in Thailand, at a seaside resort, on a tranquil day by the ocean. Suddenly, the sea mysteriously started to recede from the shoreline, exposing the bare ocean floor. Jan's husband, recognizing the pattern as the first stage of a tsunami, shouted, "We've got to run up the hill as fast as we can, right now!"

They ran and ran, getting to higher ground as the giant wave overcame them. Jan clung to a tree to keep from being washed out to sea as the first wave crashed over her, and when that receded, she ran even further up the hill, beyond where the waves were hitting.

A nurse by training, Jan helped the injured afterward. Her husband had lost a finger to the tsunami's force, and she attended to him. Then she went to the rescue of other survivors, treating their wounds with whatever she could find. She managed to help one man who might otherwise have bled to death and a woman who had almost lost a leg. Some people were beyond help.

Sometime later, I heard about this catastrophe when Jan called me from Thailand to talk with me about what she thought might be symptoms of PTSD, as she was absorbing the shock of the tragedy she had endured and the horrors she had witnessed around her.

After a reflective pause, she told me, "Being able to aid others allowed me to feel less helpless in the face of this devastation."

If people feel there is something they can do in a catastrophe, some control they can have, however small, they fare better emotionally than do those who feel utterly helpless.

"The space of awareness is small, so our personal distress looms large," says the Dalai Lama. "But the moment you think of helping others, the mind expands and our own problems seem smaller." Compassion and helping others can shift us into the secure-mode range, just as it seemed to make a difference in how Jan coped with such a devastating experience.

Even challenging events seem at least partially controllable, if only in managing your reactions to them. One of the most fundamental secure-mode core beliefs is the expectation that others will respond to our needs and help us manage our distress. People whose childhoods gave them lots of such secure-base moments are more likely to have this out-look. But even if we lacked those emotional supports as kids, it's never too late to cultivate our inner secure base. We needn't rely on loved ones for our secure mode; we can choose it for ourselves. There are many ways we can access this mode, whether through emotional connections or within our own minds.

One way we can shift out of maladaptive modes is by enhancing more constructive ones.[8] In treating depression, for instance, cognitive therapists have had some success "prescribing" activities that give people a direct experience of mastery or pleasure, both of which are suppressed in depression. This gives them a fresh perspective, if only a glimpse, which is larger than the small world defined by their negative thought patterns.

In the same way, we can prime our secure mode by cultivating its qualities. Hope and positive energy can turn challenges into triumphs—our attitude helps to shape the meaning in our lives. Sustained interest, one-pointed attention, reflection, insight, and warmheartedness help reveal meanings in our life experiences.

Reframing events that set off worry can be a way to shift toward the inner secure mode. Reframing takes many forms, as we saw in chapter 11. We might, for example, remember, *Maybe this is not about me.* It just takes a mindful pause to wake us from the pull of what we find distress-ing and thus help us remember to apply the antidote: look for the posi-tive parts of what's going on, not just focus on the negative. The good news is that these positives are always there, everywhere, even if they don't end up in the daily headlines.

At a subtler level, when we self-reference—that is, when we inter-pret what's happening to be centered on us—we often distort our per-ceptions. Though as the Thai teacher Achhan Chha put it, "No self, no problem."

Energetic Flows

Bob and Sandhi invite me to join them in the world of now, where Sandhi lives—and where Bob likes to be also (I jokingly call him Mr. Natural). But one day just before our lesson a disturbing encounter had triggered my anxious mode. I was preoccupied and a bit flustered, and I found it hard to focus.

At one point Bob asked me to direct Sandhi to run around the ring, slightly moving the left part of the rope to move her forward and the tail of the rope to keep her on the path. Finding it a bit hard to focus, I repeatedly swung the back part of the rope in her direction, at which Bob exclaimed lightheartedly, "Don't yell at her!"

I was amused by his comment on my too-aggressive use of the rope. Then he explained how my emotions had gotten in the way (although I hadn't mentioned to him my being upset) and that my use of the rope could seem predatory in the eyes of a horse.

I took all of this as a reminder to take a mindful pause and sense what was needed in the moment. Meanwhile Sandhi was waiting patiently and willingly for me to join up and play the training game. So I paid closer attention and the rest of the lesson went smoothly.

Afterward I was telling Bob about a related Tibetan concept, "speedy wind," or *lung* (pronounced "loong"), which is a subtle flow of energy through the body that reflects one's state of mind. Speedy energy (like from the anxious mode) gets trapped in the upper parts of the body, and we get agitated and preoccupied. When this happens, we need to direct the energy down to the abdomen, which calms the body and focuses the mind.[9]

I told Bob, "I think that just happened to me in this session. The nervous speedy energy was distracting me from the present. But through focusing, and Sandhi running off steam for both of us, I felt my agitated energy being released and directed into my solar plexus, just as I've found can happen with qigong."

"The same thing applies in this work," Bob said, "Your energy is directed down to the core."

While we all have *lung*, it can surge sometimes, and some people tend to have more than others. We react in different ways to this energy when it gets out of balance; the various models of maladaptive modes we've surveyed each describe expressions of excessive *lung*, which fuels a mode's underlying neurotic energy.

But these same energetic flows have a positive aspect, *lungta*, which can foster a state of mind that expresses our full vitality and life force, an expression of secure mode. Several Eastern healing systems relate such subtle energy flows to our states of mind. I've found that a qigong routine first thing in the morning helps me activate and anchor my secure mode.

The benefits can be biological. A surgeon tells of operating on a woman who had just finished a month-long intensive mindfulness retreat, a strong secure-mode enhancer. He was intrigued to see how the tissue around where he operated was remarkably supple, how it bled very little, and that the woman woke up from the anesthetic smiling!

Cultivating Lovingkindness

Compassion, the Dalai Lama clarifies, is the wish for others to be free from suffering, while lovingkindness means wanting them to be happy. Traditionally in Tibetan Buddhism, lovingkindness and other tranquility-inducing methods are seen as healing emotional obscurations, while insight and other wisdom practices clear cognitive confusions.

Any meditation practice that brings us a deep sense of calm and clarity allows us to create the sense of an inner secure base, but lovingkindness practice does this with particular power. While ordinary thoughts of our loved ones can boost us into a secure mode for a while, a more lasting change comes from intentional cultivation of the warmth of compassion.

In Richard Davidson's lab at the University of Wisconsin volunteers were taught to practice a meditation on lovingkindness. Every day they generated loving feelings toward several kinds of people: a loved one, a friend, a stranger, and a "difficult" person. Brain scans before and after two weeks of this practice showed that the key parts of their brains associated with positive feelings became more active.[10] As the Dalai Lama has said, the first beneficiary of compassion is the one who has the feeling.

Barbara Fredrickson's research shows that daily practice of lovingkindness meditation increases a person's feelings of warm social connectedness, a sign of the secure mode.[11] At the Mind and Life dialogue in Dharamsala, where I met Phillip Shaver, he proposed that practicing lovingkindness toward oneself and others would be healing for the insecure styles of being. He was struck by how similar this practice is to the security primes, like imagining or seeing the photo of a loved one.[12]

The lovingkindness practice is often done at the very end of a session of meditation, as a way of dedicating the fruits of practice to the greater good. It uses phrases like *may I be happy, may I be safe, may I be healthy, may I be free from suffering,* which you repeat silently and sincerely to yourself. Then you wish the same for people you love, for people in the surrounding area, for "difficult" people in your life, and finally for everyone everywhere.

I've adapted the traditional lovingkindness meditation here as a generic mode corrective.[13] This practice not only naturally lets our inner strengths and virtues blossom—caring and compassionate concern—but can also be done in ways that are reparative for specific modes, removing inner obstacles to authenticity and wise love.

This can be done as a practice in itself or at the end of a daily meditation sitting, or simply as an intention to wish others well that you carry throughout the day.

You can tailor the wording of lovingkindness phrases to send messages to yourself and others that are reparative for the specific needs of relevant modes, making sincere wishes that are reparative.

Make this practice your own; play with the phrases to find those that feel most fitting. Send reparative wishes to yourself, then gradually extend them to others, and finally extend them to the world at large.

Examples of some of these phrases are:

May I accept myself as I am.

May I be safe and feel secure.

May I be patient, tolerant, and kind.

May I be calm and clear.

May I be free from anxious clinging.

May I be free from avoidance.

May I experience this moment as genuinely as possible, beyond pretense, beyond defense.

May negative modes subside.

May positive modes increase.

May I be free from suffering and the root causes of suffering: attachment, aversion, and bewilderment.

May I have happiness and the causes of happiness: kindness, compassion, and insight.

This work takes great heart. Integrating a loving awareness with mindful habit change generates compassion for ourselves and for others, which helps melt the grip of our modes.

CHAPTER

13

Training the Mind

There once was a meeting of two great meditation masters, each a leading teacher in a different Buddhist tradition. Yongey Mingyur Rinpoche is a lama in the Kagyu school of Tibetan Vajrayana, and Sayadaw U Tejaniya teaches in the Theravada tradition practiced in Burma. This was a rare event; meditation masters from the Theravada and Vajrayana paths in Buddhism have few opportunities to get together.

Coincidentally, each of these masters in their younger years had undergone challenging emotional crises that they both say profoundly influenced their spiritual studies. Their emotional difficulties were clinical modes familiar in the West: depression for U Tejaniya and panic attacks for Mingyur Rinpoche.[1]

Although U Tejaniya had begun the occasional practice of meditation when he was fourteen, his mind had been flooded then with what he called the "ugly stuff." That bout in his teenage years and a second one later passed, but both times his recovery didn't last long. By the third round in his late twenties, the depression had returned with such force it lasted three years. At this point he was running a shop in Rangoon and depression tormented him, often plunging him into utter gloom.[2]

U Tejaniya decided to become ordained as a monk for a short while, a common practice in Burma. After he met the abbot of his monastery, he started to meditate again. But this time he added something new: wise discernment. He made his depression the target of a keen interest. He watched it continually, to learn how it worked. He observed what thoughts came to his mind and how those thoughts worsened or weakened the depressive feelings.

As he recalls, "Before that I'd been at the depression's mercy, but I learned I could actually do something." He could investigate and develop insight into his depression.

As U Tejaniya explains, with the application of interest and investigation comes wisdom. During his rounds of depression he had tried to meditate, exerting effort alone in a battle with his own feelings, but it had resulted in strain and tension. This time he brought another attitude: "Effort with wisdom is a healthy desire to know and understand whatever arises, without any preference for the outcome."

His depression started to lift, and the break from his negative thinking let him see the depression as impermanent; he realized he could be free of it. That made him an even more serious practitioner and led to his decision to remain a monk. And he eventually became a remarkable teacher.

Mingyur's panic attacks began in childhood as bouts of extreme anxiety. Even after he started a three-year retreat at age thirteen, he felt intense anxiety the whole first year. When the retreatants would gather daily for chanting, the clang of drums and blaring of horns "would drive me crazy," he remembers. "Panic followed me like a shadow."

His anxiety symptoms ranged from physical tension and tightness in his throat to dizziness and waves of panic. By the end of that first year on retreat, Mingyur was so miserable that he had the choice of spending the next two years hiding in his room or facing his own mind.

So Mingyur started applying what he had been taught, including using his anxiety itself as the object of his meditation. He started to see how his thoughts and rocky feelings came and went and how, like clouds passing through the sky, they lacked any lasting solidity.

He started to see "how fixating on small problems had turned them into big ones." The thoughts that had once seemed terrifying to him were now transient and feeble. His panic dissolved.

Depression and panic disorder are intense clinical modes—states of mind that take us over, but which we can also escape from. They disappeared as Mingyur and U Tejaniya each connected with a sense of peace and a wise heart.

For these meditation masters the spiritual path was not just a symbolic refuge, but also a practical one, offering emotional solace after years of pain. And their own bouts of suffering seem to have allowed each to connect to a great source of wisdom, which also makes their teaching more accessible for Western students, who so often come to contemplative practice seeking help with their emotional suffering.

U Tejaniya and Mingyur Rinpoche each found their doors to freedom in the practice of their respective spiritual traditions. Such practices are themselves ways to shift into the positive range.

There's an important connection between whatever mental skills we hone during the meditation session and how we apply those skills post-meditation, in the thick of life. As Khandro Tseringma puts it, "Transforming the mind can be practiced anywhere, anytime, and in any circumstance."

Meditation practice lets us strengthen our mental capacities. Then we're better able to apply what we have realized during sessions in the everyday world of mode reactions. If we retain the tone of the meditation experience afterward, when we are confronted with challenges we can more readily bring that awareness to bear—for instance, as equanimity in the face of our usual triggers.

"Mindfulness has changed the way I see the world and given me tools to understand my habit patterns," one student of the practice said. "Even though I sometimes get stuck in reactivity, I have faith in the possibility of seeing things clearly. Even though doubt and fear, likes and dislikes, judgments, despair, and greed continue to arise in my mind, I am willing to face them, willing to be patient, and willing to keep exploring."

To use mindfulness that way, it helps to build your capacity by practicing the method daily and, if possible, perhaps going on a longer retreat yearly, to strengthen the capacity to be present rather than absent.

In other words, practice a lot on the cushion so you can be mindful when you need it the most: in the midst of life's chaos.

Mindfulness is one of many mind-training methods used for thousands of years in Buddhist practice. While I respect the power and authenticity of these methods in their traditional context, here I'm adapting these practices and principles, extracting them from that context to share their benefits with anyone who might be helped.

You don't have to be a Buddhist to apply these methods in mode work—these are universally useful insights and methods. Applying traditional techniques in a nontraditional way yields methods that can help free you from the grip of maladaptive modes.

Calming the Mind

The old way horse trainers made a wild horse obey was by using force, fear, and even ropes to control it. The horses' instincts as prey animals led them to resist in terror, to the point of exhaustion; then they were considered "broken."

In meditation and psychological work, I've seen some people adopt an attitude like this. They are trying to control their own minds or using a "just get over it" approach. But beating ourselves up with a judgmental mind is like trying to break a horse.

We can't force ourselves to be free; we need to make peace with ourselves. Rather than a predator-like attitude—battling to force our minds to obey some ideal—mind whispering entails collaborating through awareness with our minds' workings.

Rather than breaking it in, by bringing kindness and sensitive attunement to bear we can give the mind a wide pasture to run free, letting it relax. Instead of using the force of will, we can open our mind to a will to be free.

The Tibetan word most often used for meditation practice literally means to "get familiar"—that is, to repeat the practice so often that those mental routines become habits. Meditation teaches the mind a new set of triggers and responses, with rich inner rewards. In this sense meditation comes down to positive habit change: practicing these methods until they become part of our own nature.

The Oxford English Dictionary defines meditation as "placing one's attention," or concentration, and "to observe with interest," or insight.

One-pointedness, or concentrated focus, lets us collect ourselves in a composed state of mind. Investigation of modes, particularly in the realm of emotions, is more effective coupled with such steady focus. This creates an inner equilibrium, which lets us look at the workings of our modes without judging or becoming disheartened.

Samatha, literally "calm abiding," refers to practices that still the mind and strengthen concentration, our basic mental muscle. Mindfulness cultivates the capacity for insight and wise discernment along with compassion, which unfolds empathy for ourselves and for those we encounter.

These methods work together. Mindful presence sustains one-pointedness by helping us avoid losing track of what we attend to—sustaining rather than losing interest. Bob puts it this way: "I don't want to ride a horse that's mindless or worried. I want her to be engaged and interested." So, too, with our minds.

Living freer from mode patterns is the reward of habit change. We gradually feel less constricted, more at ease, more in charge of our lives, and we have clearer connections where any were lacking.

But it doesn't happen overnight. Those with the perfectionist mode often bring their hard-striving patterns to this work, as though there were a timetable they could count on. It takes patience and a gentle acceptance of our vulnerabilities and habits while we work to free ourselves from them.

One-Pointedness

Focusing on just one thing calms the mind. As your mind grows quieter and more spacious, you can begin to see yourself with greater clarity, to perceive what's genuinely there. It's like letting muddy water settle; once the mud sinks to the bottom, the water clears. One-pointedness of attention is a kind of settling of the mind, protecting the mind from distraction.

Collecting attention on one point of focus has powerful benefits for mode work. It both calms our disturbing emotions and creates more breathing space around our habitual reactions. We're less likely to head for the exit when we face the unsettling feelings that can arise when we challenge mode habits.

If you anchor your boat, it can still experience the currents of the sea, but it doesn't get carried away by them—it stays put. One-pointedness acts like an anchor for the mind.

Find a quiet time and place, and sit quietly, bringing your focus to your breathing. Don't try to control it in any way; just be fully aware of the sensations of your breath. When you concentrate on your breath, it becomes the anchor where you continually return. If you steadily stay with the breath for a while, your attention will begin to feel more stable. The more stable your concentration, the more this mental skill will transfer to being more present to sounds, images, feelings—to all of life's experiences.

You can play with this practice—try to notice when your mind wanders—then redirect your attention to what is occurring in the present. You can always use the one-pointedness on your breath as an anchor.

Try this: when you first wake up in the morning, for a few minutes direct your mind to your breath rather than just drifting with your thoughts. As thoughts come up in your mind, let them go and return your focus to your breath. See if it makes a difference in how you pay attention to other activities through the rest of your day.

One-pointed attention is the foundation for all other meditative practices. Steadiness of mind lessens the tendency to immediately jump into some other thought, so forgetting what we were in the middle of.[3]

The Sanskrit word for concentrative meditation, *samadhi*, means to put together or collect (as in collecting wood to build a fire). To see things clearly, we first need to collect our minds by becoming more focused and composed.

The insecure modes are opposites of concentration: the avoidant cannot sustain focus on what's so upsetting; the anxious gets too agitated. Then there's the mind on automatic: a distracted, dull attention that gets swept away by whatever is interesting in that moment. Our thoughts adrift, we are driven by random impulse; our minds do not stay put at all.

But in order to carry through on anything in life, we need sustained attention, the remedy for a scattered mind. Concentration helps us work with our modes in several ways. For one, being able to sustain the continuity of a focused mind builds a stable base for sustaining mindfulness.

Being absorbed in concentration generates a pleasant state via mental means alone—a secure-mode lesson that we need not depend only on external circumstances for our happiness. This, in turn, lets us loosen the hold discontent may have on our minds, making it easier to be aware.

The impact of anxiety on clarity has been described as analogous to a pot of water: if it shakes and forms eddies, we cannot see our reflection in it. Likewise, restlessness and worry stir the mind, obstructing our view of ourselves. A mind completely caught up in the agitation of a mode has little chance of gaining clear insight into the mode itself. So a calming practice, like concentration on the breath, can be a useful prior step to becoming aware, say, of the anxious mode.

Integrative practices combine several methods to change mode habits. The calming effects of concentration were lifesavers for Becky, a client who had panic disorder, an extreme form of the anxious mode. Even going outside her apartment was stressful for Becky. She would immediately catastrophize, assuming that she wasn't safe, that something—anything—could happen. She wasn't yet a prisoner in her apartment, but she was heading in that direction.

In our sessions, Becky learned to use her breath as an anchor for mindfulness. She found that focusing on the natural flow of her breath directly calmed her anxious ruminations and allowed her to feel safe. From this inner island of centeredness, she was able to challenge distorted thinking, especially the assumption that something bad was about to happen to her.

Becky did so with heart, understanding how experiences earlier in her life had contributed to her anxiety. Her parents had been overly anxious about safety and overreacted to minor mishaps, creating an atmosphere of vulnerability. If the news reported a crane falling on a passerby at a construction site, her parents would worry that this would happen to someone in the family.

As Becky's attention became more stable on her breath, with practice, she gradually was able to rethink and reframe her distorted mode assumptions. She would remind herself of the incidents that had imprinted a sense of fear in her mind. She realized that her fears had to do with events in the past, that they did not mean she was going to die if she left her apartment building.

This inner work let Becky challenge her fearful habits and take small steps toward freeing herself from them. At first she would just walk to the outside door of her building, calming her mind by focusing on her breath, and mindfully noticing her fearful thoughts and then challenging them. When she felt ready, the next step was to walk along the street outside her building, and the relief of walking along calmly in the fresh air, as well as giving more power to her mind than to her fears, was an immediate reward.

Becky applied several transformative practices: one-pointedness, to calm down, prime a secure mode, and clarify attention; investigative insight, to view things more clearly; and compassion, for the distorted perspectives of the mode. While the essential first step in Becky's mindful habit change was the calming effect of her breath meditation, compassion played a role in two ways. For one, she was gentle with herself, not forcing a next step until she was ready. Second, she recognized with understanding the influences in her life that had contributed to this paralyzing mode in the first place.

Mindfulness Training

In insight practice—in contrast to one-pointed concentration—the specific object of attention is not really important; maintaining the observing mind is what matters. Anything that comes into awareness is the "right" target of attention. It's the attitude we bring to practice that matters.

We pay attention to the mind in a special way. Say we're being mindful of an aversive moment. We notice the aversion, but along with this awareness, there's the wisdom of discernment in the background. Weeding a garden so we can see the flowers more clearly equates to clearing away our maladaptive modes to give adaptive modes more room in our minds.

Awareness helps so much with mode work because it brings a spacious equanimity to the moment, letting us assess the situation without reactivity before making further interventions. The classical texts call these interventions "right effort"—doing whatever may be needed to keep the mind on track, such as exerting more effort if you're getting drowsy or relaxing if you're tensing up.

Sati, the Pali word for mindfulness, means "not forgetting." We observe what goes on in the mind and, when we add the steps of remembering the instructions and making a wise choice, we understand what has been so clearly observed as well as what is needed to make the mind more fit.

In mode work this means both monitoring and steering the mind as we get to know our emotional habits. Mindful monitoring acts like a security guard who vigilantly notices everyone who goes through the door, checking them as they pass. Steering the mind goes a step further, like a guard directing each person toward a destination in the building and refusing entry to those who do not belong, or receiving a parcel from a messenger, as needed.

Concentration gives us staying power. By itself it blocks insights, but when combined with mindfulness, this staying power helps us look at our emotions, especially when they are unpleasant, intense, or painful.

Insight with one-pointedness lets us calm down and sustain enough attention that we are able to look more deeply into our experiences in the present moment. This allows more room in the crowded mind, a breathing space from our conditioning.[4]

Adding mindfulness to concentration fosters a calm clarity, which can allow mode insights that might lead to reframing a situation. For instance, if you are about to challenge a distorted fear, calming yourself first can let you more readily see how the fear is in your mind, not in the actual situation.

Bringing these mind-training practices to mode work, such as one-pointedness, tranquility, investigation, insight, and compassion, helps create mindful mode work. For example, just as Becky, who suffered from panic attacks, used mindful breathing—along with challenging her distorted thoughts—to heal her disorder, integrating a calming practice with an awareness of a triggered mode can be a life raft.

It helps to practice meditation every day, and in a quiet time, so we can call on these mental abilities in the troubled moments of our lives, when we are overwhelmed or distracted and more easily give in to our knee-jerk mode reactions.

We can use these methods to handle what otherwise might become a mode trigger. Another client described taking a mindful pause when her child was in the throes of a frustrated meltdown. "I was feeling I was on the verge of a tantrum myself. But I just took a few breaths with as much awareness as I could muster, and then I was able to find a bit more presence and be clear and emotionally available to my daughter."

A windy day is unfavorable for snorkeling: the seas are turbulent, and the water is so agitated and murky that you can't see the beauty beneath clearly. It's the same with the mind caught up in a mode: when disturbing thoughts and feelings roil our minds and stir up negative reactions, we're not able to see our feelings clearly or discern what is needed.

Identifying Feelings

A meditator on a long vipassana retreat observed greed arising in her mind during lunch, when she saw a luscious-looking strawberry on a serving platter. As soon as she noticed the greed, she deliberately did not act on it—she didn't take the strawberry.

This mental sequence—seeing the strawberry, noticing greed, and purposely not reaching for the juicy fruit—happened three times in a matter of seconds. Then she moved on to other dishes. She felt very satisfied with herself for not taking the strawberry.

A while later she made a precise report of her willpower to her teacher, U Tejaniya. His response surprised her: "What? You didn't take the strawberry? But you clearly saw the greed. Awareness transforms greed, so there was no problem if you ate it."

Of course we need restraint with more harmful impulses, but he was making a different point: the specific experience itself matters less than how well we observe how our minds react to it. This approach to mindfulness was what proved so helpful in freeing U Tejaniya from the thick mode of depression.[5] And it applies directly to monitoring our modes.

This style of lightly monitoring our emotional reactions in daily life—recognizing them in passing without going into details—brings a more subtle level of attunement to our modes. While the repetitive thoughts of our modes are akin to voices in the mind, the more subtle feelings associated with the mode are like whispers.

Vedana, a Pali word, means to know or to feel. *Vedana* insight practice is one of the classic forms of mindfulness, in which we keep our focus at the level of mere sensations, without getting involved with thoughts about them. This contemplation of feeling lets us note whether the present moment's experience has been tinged with a sense of pleasantness or unpleasantness.

In a poetic passage, the Buddha likens the fleeting nature of our feelings to winds shifting in the sky, blowing from one way in this moment, from another in the next. The feelings that breeze through the mind are a bit like the wind, which can be cool or warm, moist or dry, fresh or dusty. The main quality to note with feelings is quite simple: are they pleasant, unpleasant, or neutral?

This kind of mindfulness means noticing the subtle positive or negative sensations that accompany any thoughts, images, or outright emotions arising in our minds. The content of our thoughts does not matter but rather the feelings that go along with them.

With precision of awareness, contemplating feelings means recognizing them with such immediacy and clarity that they come into our awareness before the subsequent onset of reactions, interpretations, or projections. But even if we only tune in to our feelings this way once we are under the influence of a mode, that very act can create a space in our minds.

This method gets pursued with greatest intensity when people are on an extended meditation retreat. But with clear intention we can also recognize feelings in this way in the midst of our lives.

Feelings are easier to notice than is the undercurrent of thought. Mindfulness of feelings offers a model for letting us catch a negative mode in its earliest, most subtle forms, before it has become an all-consuming mental trap.

We can develop the habit of attuning at this level by practicing with subtle feelings that are always present, by noticing the pleasantness or unpleasantness of how we're perceiving any experience. The point is not to change or fix what we notice but to be present to whatever occurs.

So when we observe with mindfulness some discomfort, we are not trying to lessen it or make it disappear. We merely observe our mental reactions to these unpleasant feelings. This lets us understand the connection between our thoughts, our feelings, and the physical sensations themselves. Our discomforts are not problems in themselves; they are just sensations. The problems come with our emotional reactions to them.

Mindful presence of the link between what comes into our senses and the likes and dislikes that arise in reaction lets us detect the underlying primal patterns: either avoidance, attachment, or indifference. Mindful presence helps us stay awake when there's a strong urge to fall asleep to our patterns.

You can experiment with this variety of mindfulness from the moment you wake up by forming the intention to notice your emotional reactions as many times as possible during the day.[6] From time to time, you'll find, a reminder to be aware will pop into your mind and you will tune in to whatever emotions you are having at that moment. Remember the basic instruction: whatever you may be seeing, hearing, feeling or thinking, notice the emotions that go along with it. The more you remember this simple instruction, the more chances you'll have to try this out.

The most basic level of emotion is simply liking or disliking. If there are no obvious strong emotions to notice at a given moment, just tune in to your sensory experience and be aware of whether you like or dislike it. You may recognize attachment, aversion, or bewilderment, hope or disappointment, or any of myriad feelings. Whatever your feelings, just note your reactions.

Relax. Accept your experience as it is, whatever it is.

You may notice a wandering mind or distracting sounds. If these disturb you, notice the feeling tone of your attitude, *I don't like this.*

Recognize and be mindfully present whatever happens. If your mind becomes clouded by the wish for things to last or to go away, just be aware of these subtle forms of attachment and aversion.

Every experience throughout the day offers a learning opportunity to notice whether the mind accepts things the way they are or judges with liking or disliking.

This can be practiced anywhere, anytime we recall the instruction. Just remind yourself from time to time to be aware.

When we observe our reactions this way, we may recognize the familiar flavor of a mode. If so, this calm observation of what's going on creates a space to reflect and to make more sensible decisions—to choose more wisely what we do.

The Lama and the Lotus

One day in Manhattan Mingyur Rinpoche mentioned casually that for three days he had not had any hot water in his hotel room. Anyone might have complained to the manager or changed hotels. But as he told the tale of his difficulty, he just laughed and laughed.

Mingyur Rinpoche is something like a human lotus leaf: as he glides along with lightness of being, things don't stick to him—not problems, not setbacks, not inconveniences like no hot water.[7]

He teaches a method that cultivates a spacious attitude in the mind—another useful skill in mode work. He explains that the Tibetan word *shinay* translates as "calm abiding," resting the mind in calmness. He likens this to soaking in a hot bathtub at the end of a long, hard day: just let go and rest, letting your mind relax in the open present.[8]

You let whatever thoughts or feelings in your mind come and go, without blocking anything. At the same time you don't get lost in them or let your mind wander. You're just aware of whatever happens, while keeping a presence of mind that doesn't fixate on any one thought but stays present to the here and now.

The point is to break our habit of getting caught in the wanderings of our thoughts and instead stay in awareness of the possibilities of the present moment. Whatever comes and goes in your mind, you neither focus on it nor suppress it. You just observe it.[9]

Being with our minds in this open way is like recognizing that, whatever clouds might come and go, the sky itself is vast, open space.

Try this in small doses at first—just for a few seconds or minutes—then gradually extend the time. Trying this in the middle of a day as we go about our lives enhances the positive modes.

One client described how she used this approach. "I was feeling increasing indignation toward someone who I have a business relationship with and who I find controlling and selfish. I decided to give it a lot of space in my mind. I didn't focus on the anger or on the person, but on a larger awareness containing it all, giving it a lot of space.

"After about ten minutes I felt my mind expanding—I had a larger perspective while staying with the issue at hand. As my attention got clarified, I felt myself becoming less reactive. I was able to remember an important detail that would be helpful when it came time to communicate with this person. It seemed more possible to stand my ground while we talked."

We can use everyday experiences as practice opportunities by checking in on our minds in any circumstance, whatever we're feeling: when we're feeling happy or sad, tired or excited, well or ill, busy or bored. The idea is to connect with an awareness not defined by sense impressions or mental states.

If we can sustain a clear, open awareness in these mind states—not just while meditating quietly—we can practice anywhere and at any time. This strengthens our abilities to stay awake while engaged in our lives—like a lotus, floating through life instead of sinking into the mud.

Wise Heart

In a small town in the Berkshire Mountains I came across a large field surrounded by woods, which held a meticulously designed labyrinth, a smaller version of the famous one at Chartres Cathedral. This maze of stone walkways adorned with floral plantings invites you to embark on a mini-pilgrimage, a timeless, leisurely stroll.

Near the entrance, a plaque encourages you to leave your thoughts behind and be open to what you experience as you meander the maze's turns and twists. As you enter, you walk under a freestanding arbor gateway. I recalled the torii gates—gracefully curved archways you pass through to enter a Japanese temple or tea garden—that I saw while visiting the temple gardens of Kyoto. They stand as symbolic reminders of leaving behind the ordinary and stepping into the extraordinary.

Great, I thought. *This is a helpful reminder to bring to any experience. I'll give it a try.* And I stepped under the labyrinth's gateway, resolving to leave behind my interpretive thoughts or any judgments about the experience I was about to set out on . . . or so I thought.

I stepped onto this mysterious path to unfamiliar territory with an open mind and marveled at the beauty of the cascading plantings that adorned the carefully placed stones aligning the pathway. I delighted in this experience. But then my gaze strayed further on down the path, which wound around and about in a confusing maze. The thought occurred to me, *I feel a bit trapped in this lovely maze.* I couldn't help but wonder what was ahead and how long it would take me to wind through this matrix.

No wonder. A labyrinth's very design triggers that wonderment, taking you to and fro, so you walk a long ways without seeming to get any closer to the goal. Just to walk a maze represents a pilgrimage, a journey that, like life itself, takes you here and there on your way toward awakening.

My thoughts wandered back to the plaque at the entrance with its reminder to leave your thoughts behind and be open to this experience. My likes and dislikes settled into the dust beneath my feet as I found myself being mindfully present while walking, without hurrying to some destination, and I relaxed into the bare sense of gliding through the maze, appreciating it just as it was. Thoughts and feelings came and went, and the labyrinth gently unfolded before me.

The late afternoon sun played on the massive oaks surrounding the ancient pathway. The soft light of dusk illumined the space between the branches. Shadows dappled the stone walkway, softening the edges of my attitudes, inviting me into the present.

With nowhere to go, no one to be, untangled from the mazes of a mind that constructs labyrinths within, I felt freed. Our thoughts can be a mere passing show in the mind, or they can become the core beliefs that solidify into our negative modes, claustrophobic inner spaces that hold us tightly simply because we fail to see any way out.

As Bob Marley sings in his song "Redemption:" "Emancipate yourselves from mental slavery. None but ourselves can free our mind."

The search for that very freedom is one reason that labyrinths were commonplace in medieval cathedrals, like the one at Chartres. These spiritual tools were in use thousands of years ago, as outward metaphors for the inner quest. A labyrinth can be seen as one of the countless methods found in the world's religious traditions for leaving behind our ordinary states of mind and entering the mode of wise compassion.

The movement I experienced from my mundane mental state to a mode freed from likes and dislikes signaled a greater degree of internal freedom. This marks a shift into a wise heart, a step beyond the secure mode. As the world's great spiritual traditions attest, the qualities that typify this mode are the very goals of inner development.

Opening to a wise heart brings yet another phase shift in our beings: our minds are free of preoccupation and self-concern. We're not distracted from being fully present. We aren't driven by impulse, longing, or aversion. Our actions align with our inner sense of purpose and meaning, as our energy is freed up for greater goals.

We are more free from fixation on rigid ideas of how things "have to be," so we are more able to go with the natural flow of life with acceptance rather than reactivity. We feel a deep sense of wellbeing—and feel happy for no reason. Our wellbeing is independent of any particular external events—someone showing they care, say, or winning praise for our accomplishments, or financial gains. The compulsion to strive for status, power, wealth, or the like subsides.

Emotions evolve within this mode. If life brings us distress, we can acknowledge feelings of sadness or anxiety but without excessively clinging to or avoiding these upsetting emotions. Resilience in the face of challenges takes the form of not losing our inner equilibrium while we stay open to a larger perspective (compared to resilience in the secure mode, where we react to stress but recover quickly).

Buddhist psychology details what our minds experience while in this mode of equanimity, with a list of specifics. The central mental quality is insight, in the sense of clear perception, along with mindfulness, which steadies this clarity.

This mode expresses wisdom and love and sees a full flowering of capacities, such as mindful presence, confident composure, gracious modesty, insightful clarity, equanimity, and compassion. Among the other notable qualities are buoyancy and pliancy in both mind and body, along with heightened adaptability. The net effect: a sense of natural ease, looseness, and joy, as well as a happiness that is not dependent on external circumstances.

The Tibetan teacher Chögyam Trungpa Rinpoche captured the essence of this mode when he used the term "windhorse" for the strength and energy that comes from connecting back to what he calls "the original feeling of being truly yourself." He calls it a source of tremendous vitality within "an atmosphere of sacredness," of "recharging your energy," where you feel you are "leading your life in the fullest sense."[1]

Intriguing signs of such deep immersion in the moment include times when we feel no boredom or no need to be entertained, or no need to soothe some vague uneasiness by losing ourselves, say, in meaningless pursuits. Our world is full in itself. Even repetitive work can be engrossing; every minute is a new chapter, a source of raw delight. Life seems to be constantly moving forward in the most satisfying way.

Spiritual traditions around the world—including in the West, from the Greek Stoics on—have long valued this. Christian monks and nuns, like Hindus and other practitioners, all follow maps of inner growth that rely on weakening worldly clinging in order to attain a beneficial mode of being which goes beyond ordinary psychological adjustment.

While my own way has been within Buddhist traditions, many different paths help us inch toward the mode of wisdom and compassion. The world's spiritual paths are rich menus of methods for the needed phase transitions in consciousness.

Most modes stem indirectly from evolutionary forces that have helped us survive. This compassionate wisdom breaks new ground, expanding the range of human capacities beyond what psychology sees and opening us to a transcendental range.

Some theorists have reconciled Eastern and Western psychology's models of wellbeing with the paradox, "You've got to have a self before you can give it up." Or as one put it, "going to pieces without falling apart."[2]

As I see it, rather than having a solidified sense of ourselves, this means having a strong confidence in our true nature and knowing how to free our own minds. In other words, a secure mode offers a platform for the next stage of inner work, where our attachments diminish.

Still, as we mature there are many lines of development: biological, emotional, social, intellectual, and spiritual. We don't always progress at the same rate along each of these lines. We can be advanced spiritually, for instance, but less so when it comes to our social, emotional, or physical development.

Tibetans use the term "old dogs" to refer to seasoned meditators who have the outward appearance of deep practitioners but lack the inner fruits. Some old dogs might be well advanced in their meditation practice but still stuck from time to time in habitual patterns that create difficulties in their lives or relationships—and so they might benefit from a dose of the methods for freedom from lingering toxic modes.

Meditative Arts

We are a week into April here in New England, and a fresh blanket of snow covers the ground. This morning I find myself feeling an eagerness for the sunlight that will spark the unfolding of nature's new life. But a reflective pause brings me back to appreciate the moment just as it is.

A memory: a spring retreat a decade ago in a Zen monastery, remote in the mountains, with a grey mist hovering in the mountain valley—and my mind unexpectedly opening to an inner source of light:

> *Awareness rests, sustaining the moment just as it is.*
> *The need to rearrange circumstances*
> *in any way whatsoever*
> *to create some semblance of contentment*
> *falls away.*
> *Attention is drawn instead*
> *to the mind's inner radiance.*
> *Fresh snow softly illumines pine branches.*

I was inspired by the centuries-old Cold Mountain poems of Hanshan, a Zen poet whose writing flows from and captures a contemplative state embodied in a seasonal moment, like a spring snow. Such reflections come most naturally when we give ourselves what Ralph Waldo Emerson called "the leisure to grow wise."

The arts can be harnessed to awaken a sense of the transformation of perception that signifies the wise heart. Meditative states that help us enter this mode have been expressed through the arts for centuries.

Much Buddhist art expresses equanimity and tranquility, or the poignancy of change and impermanence, or the direct seeing that enlivens nature poetry.

As a student of Japanese flower arranging and tea ceremony, I've witnessed this transcendental transmission, where an inner lightness draws people into a shared tranquil presence.

The Buddha spoke of the happiness and delight that flows from a wholesome state of mind, which manifests as cheerfulness, joy, and a sense of delight—"living at ease and unruffled," as a text from the time puts it. His followers often expressed this happiness in poetic descriptions of natural beauty.

A potter, who had heard me talk about meditative arts, was sitting in his teahouse when he saw an awe-inspiring view of the northern lights over a nearby lake. He made a tea bowl for me and wrote the story of the bowl. Here's how it ends:

> In that moment of awe and wonder in the dark, cold of winter, the potter had a searing insight. As he sat in stillness, he realized that just like the surface of the lake, if he could let his "heart" and "monkey mind" be untroubled and quiet, then like an inner witnessing, he could let the glories of the heavens dance through him into the earth. The heaven and earth could be one through his conscious presence and being in the world.
>
> He had a deep gut feeling that as he practiced his daily meditation and brought quiet mindfulness into everything he did—and that over time he could bring more of the love and compassion he associated with the heavens into even his most mundane daily actions and all of his relationships.
>
> And more, as he persisted in this way of being in the world, the dark parts of his own consciousness could come to light as lessons and challenges, so all of his wounds could be healed. Just like the perfect bowl, carefully shaped and fired to intensity, he could be made a worthy vessel of service in the world.

Six Signs of Wise Compassion

When I noticed the old man sitting on a large rock in Manhattan's Central Park, he had large black garbage bags piled around him. I was struck by his unkempt beard and long, grey dreadlocks. Then I saw the pigeons surrounding him in a flock, comfortably milling about near him, as if reassured by his presence.

He took some jugs of water out of one of his garbage bags and walked around the rock, pouring water into its pits and crevices for the birds to drink. Then he reached into another bag and pulled out loaves of bread. He broke them into pieces he then scattered around for the birds, occasionally taking a bite for himself in a one-for-me, most-for-you manner.

The birds fluttered around him, each settling on the rock somewhere close to him. They appeared relaxed and content together, as though he were their parent come to feed them. He stroked his beard and occasionally puffed on a cigarette while he watched the birds enjoy their feast. After a while he took more bread out of his plastic bag and tossed more pieces around to his flock.

With his long dreadlocks the old man reminded me of *saddhus* I had seen in India, wandering yogis whose dedication to the spiritual path includes taking only the bare minimum of what they need and sharing any extra with those who can use it. Such bare-bones generosity is one of the marks of wise compassion, where the reward of making others happy is sufficient.

Martin Luther King Jr. pointed out that in the biblical parable of the Good Samaritan, who stopped by the side of the road to help a stranger in dire need while others passed by, those who did *not* help asked, *If I stop what will happen to me?* The Good Samaritan reversed the question, asking, *If I do not stop to help this man what will happen to him?*

How can we track progress on our paths to wisdom and compassion? Buddhism describes this transformation in terms of six *paramitas*, or perfections, personal qualities that strengthen as we "cross over to the other shore," the literal meaning of the Sanskrit word *paramita*. That crossing over occurs in the shift from our everyday modes to the realm of inner wisdom and compassion. The six qualities of this mode are each seen as expressions of our true nature.[3]

Generosity, the first quality, reflects an openness of heart; the quality of giving is free from any expectation of getting something in return. The test of selfless giving lies not in how much or what we give but rather our motivation: our motive should be simply a concern for the wellbeing of the other person. The gift may have nothing to do with anything material; it can simply be in paying full attention or in the readiness to help someone. Generosity represents an antidote to the greed mode by giving freely, whether with our time, energy, possessions, kindness, or money.

An impeccable sense of *ethics* stems from self-discipline and a sense of integrity. The commitment to do no harm, even in our thoughts, and to be virtuous in what we do frees us from negativity within—particularly from anger and hostility, but also from remorse or guilt. With nothing to hide, we can be at ease and move through the world with a natural confidence.

What we think of as *patience* emanates from a deep quality that also gives rise to tolerance and acceptance. These qualities are based in an underlying strength that lets us keep our composure in the face of difficulties. Such inner tranquility does not reflect a suppression or denial of feelings but comes from a genuine openness of heart, in combination with insight into the transitory nature of things.

An enthusiastic *vitality* lets us pursue our spiritual goals with joyful perseverance. From this joy comes the courage to endure hardships whenever necessary and the vigor to sustain the heightened levels of practice the mode requires. Much of this energy comes from realizing the benefit we can be to others.

Strong *concentration* is at the heart of mental stability. While in a distracted, daydreamy state, we run through the same habitual modes over and over. Training our minds to focus allows us to do what we want them to do rather than be blown by the winds of circumstance. As we'll see, such presence of mind can open the door to mode shifts.

The sixth *paramita* unfolds as we access the wise heart's *wisdom* level, and the other five *paramitas* integrate with this wisdom. Here we gain an awareness that transcends words, thoughts, and concepts of any kind, including the distorted lenses of modes that veil such crystal clear perception.

That man feeding the pigeons in the park reminded me of what my teacher Tulku Urgyen Rinpoche called "hidden yogis," people out working the fields or cooking while maintaining high levels of meditative practice. Of course I don't assume the man in the park was one, but he expressed a lot of selfless compassion.

The *paramitas* are under the umbrella of a deep humility—to show off one's virtue runs counter to their spirit. In some Japanese monasteries monks are encouraged to anonymously do good deeds for others, as secret helpers. "The back of your practice," a Zen saying goes, "is as important as the front." In other words, it's not how you look sitting on the cushion that counts, but what you do when you're off it.

Waking Up

My late teacher Adeu Rinpoche was one of those wise guides on the path who embodied lightness of being. He once spent several days telling us about the seventeen years he had spent in a Chinese concentration camp.[4] Despite the brutality and endless days of slave labor and near-starvation, he bore it with equanimity.

"Your state of mind cannot be imprisoned by others, only by yourself," he said. "The prison guards had no idea what I was thinking or imagining. I was free to remember my own root guru and do any practices. There was nothing the guards could do to control that. They could control my body but not my mind."

Such lightness of being indicates living with compassionate wisdom as one's default mode. That inner freedom signifies the lotus effect at full flower. Throughout Asia the lotus flower symbolizes a purity of mind that floats free of worldly conditions and constraints. Legend has it that when Gautama Buddha was born, he took several steps, and a lotus bloomed at each one.

The lotus leaf, an electron microscope reveals, is covered with teeny bumps, just a few microns in size (a micron is one-millionth of a meter). These eensy spikes, remember, poke up to gently meet whatever is on the leaf so that whatever contacts the leaf's surface has little to hold on to.

Water droplets roll off these soft spikes as though they are ball bearings. In the same way, our being lightens as we move through the secure mode to a wise heart and then to an awakened perspective. The lotus effect increases so that mode triggers roll off instead of hooking us, as lotus-like lightness predominates.

"As practice progresses, things don't stick to the mind," says U Vivekananda, a meditation teacher. "It has a repelling resistance power—a Teflon quality—not bothered by what's happening. The mind bounces back from negative thoughts. It doesn't want to stay there."

The 17th Karmapa puts it in terms of whether difficult situations disturb our peace of mind: "Imagine a mirror. If you put something heavy in front, the mirror will reflect it clearly, but it doesn't bear any of its weight. It is possible to experience difficulties in the manner of the mirror, without being weighed down by them."

Reflect on a caterpillar before it sheds its sheath and reveals its wings. Crawling and flying are modes of being. One being inches along, undulating as it pulls its body forward; the other glides above us, freed from downward pulls. This transformation from caterpillar to butterfly occurs within one and the same organism; shedding a sheath makes all the difference between crawling and flying. For the caterpillar, it's a matter of timing.

We, too, can either crawl along clinging to the ground or glide with a lightness of being. What allows us to connect with our inner freedom? Like a newborn bird protected by its nest from the force of gravity, how much of our own awakening depends on readiness? And how much on how we perceive?

The awakened state represents a dimension of being that modern psychology's theories of development do not map. The very idea of non-conceptual awareness is largely unknown to modern science. Psychological theory lies within the realm of ideas, while the awakened state deconstructs mental models of any kind.

Our mode hooks operate at the level of emotions. But a more subtle level of mental hook can be seen in what Tibetans call *dzinpa*, a term for fixating or holding to a thought—any thought. Ordinarily our minds offer up an endless stream of thought after thought after thought, catapulting us down one train of association after another.

But inner lightness comes with the awakened perspective, where the mind floats entirely free of such fixations. We can let thoughts come and go, as they will, without the least grasping. When we do, thoughts tend to float away. Watch how a wave rumbles through the sea, gathering momentum, temporarily concealing its transparency before dissolving back into clearer waters. Thoughts are like waves in the open sea of awareness.

This mental lightness utterly transforms our experiences, altering how we relate to whatever is happening. When the Tibetan sage Padmasambhava was asked to name the signs of great spiritual progress, his list included that destructive emotions do not arise in the mind. But a still greater sign, he said, was that "you are not directed toward any object whatsoever." In other words, attention is utterly open and free of the least attachment.[5]

The dividing line between a wise heart and full awakening lies where we free our minds from fixations in this way. If we cultivate positive qualities and loosen attachment, the path is left open for a wiser mode. But to shift into the awakened level requires yet another phase transition: freeing the mind entirely from the hold of our self-clinging thought patterns. Here, we stabilize in thought-free wakefulness.

From a Buddhist perspective, there are two truths through which to frame our life experiences: relative and ultimate. Relative truth includes our conventional understanding, personal stories, and modes. From an ultimate perspective, modes evaporate and are no more real than a moon reflected on a clear pond.

A wise heart lies within the relative realm, that of ordinary thoughts and concepts. The awakened perspective, though, allows us access to a realm beyond concepts in the domain of an ultimate reality, one free of even the mental categories that define our ordinary world of experience.

Just as a finger pointing to the moon is not the moon itself, any descriptions are at best the roughest indicators of the awakened way of being. Since the awakened state operates in a nondual realm, free of any mental category, the very attempt to put it into words already does an injustice to the truth of that experience.

In the awakened state, the mind engages with the world without the least resistance. Tulku Urgyen Rinpoche was another who embodied these wisdom qualities. He once said, "In a moment of yielding, whatever *is* comes into play." His Tibetan translator, Erik Pema Kunsang, clarifies the meaning this way: "It takes a kind of courage and compassion to yield to things as they are. Usually some insight into the nature of things comes out of that."

As our modes shift from dense to ultra-subtle, we transit into what's called "pure perception," said to be free from the distorting lenses of our everyday minds and beyond the very category "mode."

We relate to the comings and goings within our own minds with ultimate nonattachment. Our mental models are realized to be subtle forms of fixations. With this insight, what was solid vanishes into air.

This profound re-perception of perception itself distinguishes the awakened perspective from a compassionate wisdom, where our attachments are weakened but we still operate within the mode of thought. The mind of an awakened being becomes free from all fixations.

The darkness of confusion, in its forms of accepting and rejecting, attaching and avoiding—and all our hooks to the maladaptive mode known in the East as *samsara*—becomes transparent in the light of awareness. With that comes a deep wish that all beings understand the true nature of their minds.

Some say that the cluttered mind of suffering lacks space, making it hard to see our genuine nature. Exhausting the wanting mind releases the grip of the self. Within the spaciousness of awakening, thoughts lose their solidity. Projections, assumptions, and reactions fall apart.

Through this lens, how poignant the human condition is, chasing ephemeral rainbows and building castles out of sand. At the same time, compassion builds for the relentless conditioning that perpetuates *samsara*.

Take compassion. Negative modes often foster obstacles to compassion, such as self-absorption, lack of empathy, or confusion about how to help. In the secure mode we feel compassion but typically express more concern toward those we love the most.

As we mobilize this mode, our compassion can become more impartial, eventually extending to a far wider circle of humanity. And in the awakened view it takes the form called "great compassion," where we feel ongoing, spontaneous concern for the suffering that *samsara* itself creates in people everywhere.

A compassionate mirror reflects in a kind way. From true authenticity arises a gentle wisdom, which accepts people as they are and empathizes with how they perceive reality. Wise compassion holds no favorites, viewing everyone equally through a lens of love.

"Biased compassion is mixed with attachment," says the Dalai Lama. "Unbiased compassion is mixed with wisdom." This brings the effortless, spontaneous flowering of qualities, such as patience, kindness, forgiveness, self-discipline, contentment, and so on.[6]

Many people thought that the late Dilgo Khyentse Rinpoche, a revered Tibetan teacher, embodied an awakened state. He was famed for his openness to anyone and everyone. Throughout the day and long into the night he would receive a steady stream of people; he would never turn anyone away.

He once said, "The everyday practice is simply to develop a complete openness to all situations and emotions, and to all people, experiencing totally and without mental reservations and blockages, so that one never withdraws or centralizes into oneself. This produces a tremendous energy, which is usually locked up in the process of mental evasions and a general running away from life experiences."[7]

Human nature at our core being harbors a basic goodness. This essential nature is our ultimate secure base.

Teachings of Life Experience

In the early morning hours in the barn where my older horse, Bodhi, was being kept, as his life was drawing to a close, I stayed attuned to caring for him, trying to sense from moment to moment what his needs were. I stood still so he could bury his head into me as I stroked his forehead to comfort him. I sensed when to allow him the space he needed as he paced his stall to ease his discomfort.

I felt the challenge of balancing a calm, loving presence while not letting my own feelings get in the way. When you see someone you love in such discomfort, it's hard. But I had to stay clear and focused, ready to act on what was needed.

It all dissolved so fast. There was no time to prepare—just stay riveted to the present, awake to the unfolding. Looking into his eyes, I felt the deep tenderness of our love.

He began to stagger, seeming to want to avoid falling on me, then collapsed to the floor. I knelt by his side trying to comfort him. With his last bit of strength, he reached his head toward my face. I kissed his nose as I always had in our many years together, and gently held his head.

I watched in awe as the life force lifted from his eyes, revealing a luminous gaze. His eyes melted into transparency. His life force, our tender love, his sweet personality, dissolved into a clear, empty, peaceful essence. It was a moment when the relative nature of things dissolves into the greater expanse of luminous transparency.

I sat with him, silently experiencing a profound empty presence, humbled by the natural force of change. A warm glow of love filled my heart with an inner sense of his spirit.

It wasn't until the next day that the impact of losing him and the sense of personal loss hit me. It takes time to integrate the loss of the ones we've cherished. There are many layers of association, warm hugs, little habits of relating, and a love that was once defined by a personal connection, which needs now to redefine itself as enduring without form, as invisibly connected.

In the days following, heartfelt condolences poured in from caring friends and relatives, sharing the common experience of parting with someone we've held so dear. Love and compassion soften the truth of impermanence. Tibetan teacher Chökyi Nyima Rinpoche, says, "When we lose someone we love, we are shocked and see how fragile life is. But everything that happens, every instant, is fleeting, momentary change, transient. Everything is in motion. Everything changes. That we need to accept."

Our life experiences can directly reveal to us glimpses of the wisdom of universal truths. These are tastes of the perspective from a mode of wise compassion.

In the Tibetan tradition this is called the "symbolic teacher of life experience"—finding the larger meaning in the events of our lives. Everything that happens is a message, if we are open to it. We have a choice of the lens through which we perceive. How we frame our experiences makes all the difference.

In Hermann Hesse's *Siddhartha,* a story of the Buddha, the flow of a river becomes a symbol for that same wisdom teaching, the transitory nature of things: it was never the same river but constantly changing from moment to moment.

Watch a sun setting and reflect on the truth of impermanence as the golden ball disappears beneath the horizon. We can learn from anything, if we orient that way. Life's moments become reminders of a larger perspective, one that does not close in on the small world of a limited mind state.

Take a simple Facebook post from a friend about moving to a new apartment: "The room is filled with boxes, but there is still space."[8] Reflect on the meaning of "there is still space" and how it applies to everything. Even physics tells us that an atom is mostly space. A wise compassion reflects a spacious mind.

The intention to hold such larger perspectives as much as possible has a beautiful expression in a prayer:[9]

May my mind turn toward the Dharma.
May my Dharma practice become the path.
May the path clarify confusion.
May confusion dawn as wisdom.

15

The Physics of Emotion

On a small island in the French West Indies, as I sit peacefully on my balcony with a notebook, I notice a persistent mosquito alight on the page. I reflect on the narcissism of mosquitoes as I watch it try to land on a juicier offering for a tasty meal (which would be me), hoping to leave its noxious sting behind. I can feel my annoyance gradually build and fully occupy my mind's increasingly narrowed space, which had been wide open just moments before.

Swatting the damn thing isn't an option for me, since I am determined not to take life, even in its most unappealing manifestations. I could leave my balcony and hope that it doesn't follow me into my room. I could put a mosquito net down over me and hope that he won't be trapped inside with me.

Since I've already sprayed my (rapidly diminishing) paraben-free insect repellant on every square inch of my body, I even entertain the possibility of letting him bite me, with the hope he would then go away. But that seems foolhardy given that mosquitoes here sometimes carry dengue fever.

So I tiptoe further outside, hoping he doesn't notice and follow me into a place in the sun, where perhaps it will be too hot for him to go. Now I'm outside, gazing again at the open expanse of water, and I am reminded that the open, clear nature of the mind is always there, just behind these unpleasant experiences.

A rain cloud has started to gather right above my once-sunny mosquito-free fly zone. I watch how the mind again starts to look for the next place to escape. I wish for a sting-free, rain-free safe place. I should just get a grip and be grateful that I can write in such a beautiful setting.

And then the tropic's bright warm sun blasts through the looming rain clouds, and I relax . . . except when I suddenly recall those warnings about the thinning ozone layer: I realize I forgot to bring my sunblock!

On and on it goes as I observe closely this relentless pursuit, the mind recoiling from the unpleasant and yearning for the pleasant to last longer, wishing it wasn't so fragile and impermanent. Watching my mind go through these changes, I finally settle into contemplating a giant palm leaf swaying in the wind, dancing through space, and I delight in the gracefulness of the lush foliage. Through the flowing ferns I gaze once again at the open expanse of turquoise water. As my mind settles, a building wish within me for the roosters to hush their assault on my senses gives way as I quietly appreciate the symphony of sound.

Our minds run through a cascade of perceptions and reactions from moment to moment. One feeling tumbles into another in an endless flow of subtle (and not-so-subtle) reactivity. We face a constant challenge to get a persistently boisterous world under some semblance of control.

Buddhist psychology offers a helpful model for sorting through this continuous stream of thoughts, feelings, and perceptions to see how they create our reality—and our modes of being. This model is called the "chain of dependent arising," meaning that any one link depends on the previous one.

The chain of cause-and-effect links in the mind, put simply, runs from sensing to liking or disliking what we perceive. The resulting emotions and thoughts can lead to clinging or rejecting, shaping our subsequent intentions and actions. A qigong teacher says that *yee*, a Chinese word for mindfulness, also means intention.

Cognitive science tells much the same story, but in different terms. The brain takes in information from the senses and, very soon after, it goes to the emotional centers that form a judgment: like or dislike. From there a sequence unfolds that leads us to a set of thoughts, emotions, and actions—often just repetitions of habits we've played out again and again. Our modes are major sets of such habits.

I'm reminded of a comment by the Dalai Lama about how the insight into dependent arising explains "how an individual revolves within the cycle of existence." We fall into our negative modes over and over, revolving through the same old patterns.

Whenever we find ourselves in such a mode yet again, deep habits of mind are likely to take over. We see the world through that particular cloudy lens and interpret with the distorted patterns of thought that stem from that unreliable way of seeing our world.

The cause-and-effect sequences present in each moment are, for the most part, benign: the sensation of itchiness leads to feelings of unpleasantness that in turn becomes a soothing scratching of the itch.

Some are not so benign. Take a classic example: if the senses find what we contact pleasurable, a mild liking can sometimes grow to greed and unquenchable craving. Likewise, if that sequence starts with a reaction that triggers a negative mode, then we end up in a virtual prison of the mind and heart without realizing how we got there.

Our troublesome modes begin with unconscious cause-and-effect sequences, ruts in the mind that have become well-traveled emotional superhighways to our troubled realms. To the extent that they are maladaptive, such mental habits are what Buddhist psychology calls "conditioned suffering." The good news is the "conditioned" part: since we have learned these habits in the first place, they can be unlearned.[1]

The cause-and-effect in every moment of perception resembles something like the physics of emotion. The choice comes down to whether the causes and their effects reinforce our undesirable modes or intentional changes can be made to free ourselves from them.

It can become easier to observe these subtle, usually unconscious sequences on an intensive meditation retreat designed for that purpose. These are meant to lead us toward a wise compassion.

But we can also extend this awareness into daily life to detect the subtle causes—triggers—for our negative modes to launch. And the instant we bring this mindful discernment to a troubling mode, we have the potential to break the links in the chain that hold it in place.

Chocolate Mindfulness

A client who loved chocolate was starting to get hypoglycemic symptoms, like headaches, when she ate it. She decided to use her love for chocolate as a practice opportunity.

When she noticed herself wanting to eat chocolate, she instead would focus her awareness on the desire in her mind rather than the object of her desire. She became aware of the mind of wanting, which grew in intensity from a mild attachment into outright craving. She mindfully restrained her building intention to reach for chocolate and stayed attuned to the sensations in her body and her thoughts.

As she more closely observed her cravings—neither giving in to them nor turning away—she sensed a slight note of sadness beneath the other feelings. Investigating why her chocolate craving should evoke sadness brought the realization that this was tied to a mode pattern, a craving for nurturance.

She recognized feelings of emotional deprivation, a familiar theme in her life. As she continued to closely investigate the discomfort behind the sensations of craving and the sadness of this unfulfilled need, she used a mindful presence to gently hold this discomfort. She felt a loving, patient awareness grow. The unsettling feelings began to dissolve and eventually fade away, to be replaced by an increasing sense of a nurturing awareness. She saw how this inner nurturance could be both more powerful and more accessible than the nurturance she sought through her craving. Even chocolate paled by comparison to the nurturing power of awareness.

When my client investigated her attachment mode, she applied several mind whispering practices: She recognized the mode in action, tracking its distinctive thoughts, feelings, and impulses. She used restraint as an antidote to attachment. She connected with the nurturing power of awareness itself. And she put mindful habit change into practice.

Tracking the Chain

Samuel Johnson, the eighteenth-century man of letters who compiled the first English-language dictionary, was known for his witty aphorisms. Johnson once said that he always endeavored "to see things as they are, and then enquire whether we ought to complain."

An emotion begins as we appraise what we perceive. This either draws us toward or repels us (or, as Johnson would have it, "we ought to complain"). That great emotional duality—like/dislike, desire/dread, hope/fear—occurs in circuitry linked to the amygdala within the more ancient parts of the brain.[2] These decisions occur in unconscious circuitry instantaneously.

The thinking centers in our neocortex have no direct control over what we find ourselves desiring or dreading—only over how we respond to those urges once we feel them. The design of our brain means we cannot stop the continual urge to approach this or avoid that—just to choose not to act on it. But if you can acknowledge or take note of these feelings, you can undercut the subsequent projections, desires, irritations, and the like that could otherwise build into a negative mode's core beliefs. You can simply acknowledge that this is how the mind works rather than believe these judgments.

Here's how a student in one of my workshops applied mindfulness to track her subtle likes and dislikes:

In most meditations I could see aversion (also known as disliking, disgust, or hostility) . . . I see my "not liking" is so quick I hardly notice it, and most of the time I don't notice it at all. For example, it's cold in the room. I cover up my head with my sweater hood and watch inwards as I do this action. I become aware that there is a very, very, very subtle whisper of a thought going on at the same time, and I also become aware of a subtle sense of irritability.

So at that moment I put questions to make the whispers come into view, if possible: *Why am I doing this? How am I doing it? What is the cause? How do I feel? What are my thoughts?*

I feel uncomfortable is one answer. *I want to be warm. I don't like being cold* is another. So I recognize the dislike, the aversion, and I decide to watch that more closely. I see it's the dislike of cold, of being uncomfortable. I am able to see some of the steps before and after that subtle thought. My mood had begun to develop from physical discomfort, then to irritability, annoyance, and could have turned into anger if I didn't source the cause—and if I didn't have a sweater.

She was closely observing her "automatic judgments," that initial evaluation of an event as good or bad, which registers in milliseconds—ordinarily, outside of our awareness. If we leave those thoughts to run on unchecked, they carry seeds that can blossom into full-blown modes. But if we bring more awareness to them, they just blow away in the winds of our minds.

Remember the micron level of the lotus effect, where the droplet only contacts a tiny spike on the lotus leaf instead of a flat surface where it could spread out and stick. That spike is akin to the repelling quality of mindfulness.

The Bare Facts
Over the years I've done intensive meditation retreats with Sayadaw U Pandita, one of Burma's most respected masters of vipassana, the Pali word for insight meditation. Aung San Suu Kyi numbers among his students.

During retreats—lasting from several weeks to a couple of months—I got used to U Pandita's way of being intensely focused and present, with laser-like precision and investigative awareness. Sayadaw has a keen interest in science, and once when he was teaching in the Washington, DC, area we took him to the Smithsonian Institution's natural history museum.

At the Smithsonian Sayadaw brought his laser-like precision of mindful attention to investigating every single showcase. At each, whether the headgear of a Native American tribe or a brontosaurus skeleton, he would carefully inspect every detail and read all the explanations to be sure he understood it all. It took hours to get through just a few wings of the museum because he took a keen interest in *everything*, wanting to know how it worked, the laws of nature it illustrated—and he would make connections to Buddhist principles.

Then we came to a glass enclosure where two insects had started what had become a gigantic swarm of hundreds of thousands of descendants. He was utterly fascinated, while I felt utter aversion. He was examining how the bugs related, moved in patterns, and so on, while I was finding every possible way to avoid having to watch.

In mind whispering the aversive mode is to be brought under the same investigative scrutiny as any other thought or feeling. But I can't say I was very successful at examining the nature of my aversion to those crawling bugs!

"Vipassana" means seeing with wisdom in a heightened awareness of the mind itself. Concentration is like staying at home: quite peaceful and tranquil. "Vipassana," as Sayadaw U Pandita explains, "is like leaving home to go out and explore, to learn more."

Vipassana meditation, or insight practice, typically begins by building strong concentration, using the breath as a focus. Then a switch to mindfulness opens our attention to whatever passes through the mind—any thought, sound, sight, whatever it may be—and simply notes that object of awareness without getting caught up in it.

On a vipassana retreat we bring this "choiceless awareness" to whatever we experience throughout the day, not just while sitting on a cushion. Sustaining this attentive presence means, ideally, being continuously aware of every moment with accuracy and precision. After some time, mindfulness drops to another depth, penetrating deeply into the nature of awareness itself.

When I would report my meditation practice to Sayadaw, I sensed that he had mastered the map of consciousness so thoroughly that he knew exactly how to guide me; learning from such a master has been life-changing. I learned how to describe just the bare facts as precisely as I could in those reports—no adornments or creative embellishments—including any insights into cause-and-effect, impermanence, or selflessness.

The goal is to realize in our own experience what Buddhism refers to as "things as they are": the cause-and-effect interdependence that arises and passes away continuously, and the unsatisfactory sense when it comes—or goes—against our wishes. As we begin to see phenomena arise and pass away in line with cause-and-effect, we get a sense of their impersonal, "selfless" natures.

These practices are aimed toward a compassionate wisdom. But they can also be applied to changing our mode habits.

Everyday Habits

You can turn any experience into a practice opportunity.

Trying to change an everyday habit is particularly rich. Say you feel you're not drinking enough water and want to get in the habit of drinking more. From time to time throughout the day you would bring a mindful, close observation to your physical sensations. This lets you be more aware of the varying signs of dehydration, like subtle thirst. Note how the feeling of thirst leads to a desire to quench your thirst and an impulse to reach for a drink of water.

This offers a chance to observe more finely as you are changing this physical habit. You can mindfully observe the subtle nuances and cause-and-effect sequences that make up the simple act of drinking a glass of water. With close, mindful observation you can slow the sequence down to notice frame by frame the physical and mental activities involved:

You note the physical sensation of thirst . . . the mental intention to reach for a glass of water . . . the changes in your muscles and posture as you reach for the water and lift it to your lips . . . the intention to drink . . . the physical act of drinking . . . the feel of the water as you swallow . . . the signals of quenching your thirst.

As you observe closely the intentions and actions involved, you can see how our mind–body processes play out their habitual routines without a moment's thought. What had been automatic and unconscious habits come into the spotlight of awareness.

This heightened awareness can help you attune to signals of thirst that you might otherwise have ignored and so guide you to drink more water through the day. But this can also be extended to other levels of habit change, from other automatic daily routines to working with modes.

Even small awakenings help you break the trance of automatic habit. Mindful habit change applies to any mode. When you work with a negative mode, it can help to focus the habit change on specific patterns that you find most prominent within the mode.

For the anxious mode, it might be that your thoughts are going through a loop of worry again; for the perfectionist mode, that you are working longer hours than you need to; for the predator-like mode, that you are using I–It messages. In each case you would make the appropriate switch to challenge your thoughts, to calm down, to attune to another person, or to have more fun.

Beyond that, by awakening awareness more intensely these insights into the natural principles at work in our lives, and our minds, can help us see more clearly the cause-and-effect, transitory nature of an experience as simple as thirst and its quenching.

As habits of thinking and feeling, acting and interacting, our negative modes fit into what Buddhism calls "conditioned suffering," learned habits that can be altered. But these habits are tenaciously resistant to change. For starters, they are hard to see clearly because they fade into the background of who we are used to being and how we usually act.

Our mode habits reside in the bewildered, hidden aspect of our minds. A mindful awareness spotlights these habits, bringing them into focus, breaking the trance of habit. Mindfulness deconditions mode habits as we gradually connect more and more with a larger awareness and become less defined by these ways of being in the world—as we replace confusion with clarity.

Like the open expanse of sky that outlines clouds passing by, our emotional habits move through and dissolve into a larger dimension of our being. "You are the sky," says Pema Chödrön. "Everything else is just the weather."

Unhooking

Hooks refer to our readiness to activate a mode; Tibetans speak of "the charge behind our thoughts, words, and actions; the charge behind like and dislike."[3] That same charge energizes our urges, animates our moments of tightening, and drives the story lines we tell to justify what we do. It primes our negative modes and determines our readiness to become triggered.[4]

When our hooks catch us and drag us into a negative mode, we operate on automatic pilot. We get swept away in the reality the mode dictates, following the same old urges, hopes, and fears as willing creatures of habit.

The small annoyances or impulses that run through our minds as we go through our days are moments of getting hooked. Each one offers a ripe opportunity to practice *un*hooking, letting go of the charge. If we practice unhooking with the small things that catch us, we'll be better prepared to unhook from bigger things.

Awareness can prevent a hook from taking hold, before it drives us to react. Our negative modes need something on the receiving end to stick to; if there's no clinging, there's no hook—like when one side of the Velcro tape falls off. As a discerning mindfulness builds in strength, we can turn our awareness to that which is getting hooked. Awareness begins to loosen the grip.

If the full-blown mode does trigger, once we notice that we're in a moment of being hooked, and acknowledge what's going on, we still have the option to stay with what it feels like to be triggered, experiencing the thoughts, feelings, and urges to act.

At that moment, says the Tibetan teacher Dzigar Kongtrul Rinpoche, we are in touch with the essence of the human condition, the need to manage ourselves in a world that continually shifts around us. This background unease underlies our susceptibility to being hooked in the first place. Getting hooked seems to give us something to hold on to amidst the uneasiness of existence.

Where on the chain are we getting hooked? The answer suggests what might be a wiser choice. When the hook is mostly cognitive, we can work with our perceptions and thoughts. But when it is more emotional, we may benefit more from a different approach, one that can help to access the hook: empathy, acknowledgment, inquiry, or recognizing underlying memory triggers and psychological fixations.

We can train in sitting still with whatever the hook elicits in us. If we can simply label the hooking thoughts with a quiet whisper in the mind, *thinking,* and then let go of them, as they dissolve we can bring our focus back to the present moment.

Lacking mindful presence, our negative modes can make us either want to run away, hang on for dear life, or dissociate. Mindful discernment lets us interrupt our habitual chains of thought and feeling, our knee-jerk urges and reactions—eventually even during the strong modes that sometimes rule our minds.

Modes can be subtle. But in a reflective moment you might find yourself catching a thought that has the flavor of a mode and reminding yourself, *That's an aversive mode.*

And you might sustain awareness to observe the momentum of that thought's force fade away, watching it play out its cycle. As you are able to do this, you'll see you can have the thought, but this time—within awareness—you don't have to get hooked into believing it.

Awareness that recognizes what's going on can often nip the mode in the bud. As we strengthen this ability and make it a familiar habit of mind, our awareness becomes stronger than our hooks and we have the power to stop the usual chain reaction—like clicking on the DELETE key rather than on the SAVE.

Our modes are opportunities.

De-Linking

Sitting by a lake, practicing and writing, I notice fewer thoughts being generated as I watch the spreading ripples of water rise and dissolve in continuous motion. Like the ebb and flow of the rippling movements, the deep stillness of the mind frames the natural rhythm of the appearance and dissolution of my thoughts.

The temperature had dropped significantly that day and a thin layer of ice was beginning to freeze the ripples together, just as the mind solidifies patterns of thought. There was a flow of water beneath the thin sheet of forming ice, like the steady flow of background thoughts in our minds.

Like the frozen ripples, our fixating thoughts and emotions can seem so permanent. Just as water turns to ice, so the mind solidifies through its tendencies to fixate on how things appear.

Awareness has been likened to a sunlit sky. The warmth of awareness melts solidified thoughts and emotions, just as the sun melts the rippling patterns of water temporarily frozen in ice. And just as ice melts back into water, the ice of our solidified mind states can be transformed by our awareness, allowing these temporary fixations to melt into the expanse of awareness.

When it comes to a mode's frozen habits of perception and reaction, disenchantment with the way we expend our life force to fuel these relentless pursuits can bring a thaw. Being disillusioned with the states of mind that drive us to hold on to or strive to possess what does not bring contentment—or with our failure to simply appreciate our pleasures without clinging to them—can help us find contentment with things as they are.[5]

When we shake ourselves loose from the mind that clings, we free ourselves to be present and aware, bringing a fresh perspective to each moment. It's the mind, the perceiver rather than the external experiences that can bring truly lasting contentment.

We can be more appreciative of others, more attentive to the people in our lives, and more available to those in need. One way to measure progress is to ask ourselves questions like, *Do I crave less? Am I more kind? Less reactive? Less driven by disturbing emotions?*

A man in one of my workshops told me he had been addicted to a range of substances for years. Then after being introduced to this approach, his life had changed. "What had changed?" I asked him.

"Without space, when you feel that craving triggered, there's no choice. But when I felt craving, I'd give it lots of space and saw I didn't have to react. Mindfulness allowed me to step back from those feelings and be aware of my reactions—and the craving would pass."

Negative modes seem more likely to kick in when we look outside of ourselves for our needs to be met, say by some drug, or when life disappoints us, or when we exhaust ourselves trying to control our world or change other people. Trying to change outer circumstances to find inner contentment sets us up for frustration.

There's a turning point that comes when we lessen the hold of habitual ways we cling to in order to avoid our experience. Resisting the gravitational pull of our modes awakens a momentary freedom. Once we see our learned habits of mind for what they are, our identification with them can start to dissolve.

As we do this, there's a shift in the ground of our beings. We find a growing patience as our judgmental minds melt into more discerning and accepting awareness. We begin to see things a bit more free of distorting lenses, including the poignant realization of how we, and the people in our lives, can perpetuate suffering.

There can be an increasing sense of poignancy, a disillusion with our modes of conditioning and their limiting habits of how we see ourselves and others, how we react and feel. When the surrealist Magritte described the mystery of his art with the phrase "the absurd mental habits that take the place of an authentic feeling of existence," he could have been talking about our negative modes.

Such disenchantment highlights the great relief we feel as we do what it takes to free ourselves from our negative mode habits. As they become more transparent, we can loosen the grip of the self-protective patterns that have imprisoned our minds.

The Two Levels

A vipassana teacher I know well spends several months a year in a solo meditation retreat; he has a delightful lightness when he's out in the world. Once he said, "Lately I've been enjoying catching the earliest sign of 'I,' like a hint of pride."

That radar for the building blocks of self signifies one fruit of meditation: the ability to recognize the causes and effects that create our experiences. In traditional meditation practice we look into the causes of our most subtle discontents. As Khandro Tseringma says, "Self-cherishing and self-clinging are the real culprits." We can distinguish two ways these methods can be applied in the pursuit of liberation: at the relative level (for example, as I've adapted these methods for mode work) and at the ultimate level in realizing lightness of being.

Mindfulness is not in itself wisdom; rather it plays a preparatory role, setting the mind's stage to allow wisdom to arise. At the relative level, mindful inquiry can recognize our negative patterns, and so create the internal space that gives us a platform to loosen its grip. A more subtle level of vipassana, holds another possibility: allowing the knowing mind to inquire into its own ultimate nature.

Once we perceive this finely detailed level of our minds' workings, an intriguing possibility arises: we can "de-automatize" our habits of perception by bringing them into the full light of awareness.

We don't have to get rid of the waves in the ocean; we can let them dissolve back into their original nature, the vast sea. The same with thoughts and emotions; they appear and dissolve into the greater expanse of awareness.

As this steady attentiveness regards the movements of the mind itself, we can realize three truths about the mind's nature. The principles revealed by strong and sustained mindfulness and discernment are

- inconstancy, or the impermanence of all things;
- the uneasiness of existence, or the nature of suffering; and
- the "emptiness" of what we usually regard as a solid, fixed self.

These three principles can be found within physics as well (in the universal dynamic of constant change) and in cognitive science (in the deconstruction of self into mental processes). But Buddhism adds practice to theory: to experience these truths in the core of our being.

There is no single, fixed path. Very diverse traditions—here, from meditative practices to psychotherapy to principles in horse whispering—can play a part in leading us to our inner home, clearing emotional and cognitive patterns that obscure our true nature.

While psychology's mode theories focus on how our mental fixations warp our perceptions and relationships, the East zeroes in on our ultimate relationship: the one between our consciousness and our very thoughts. Practitioners become aware of the process that builds our mental models of ourselves and of our world and recognize that they are mere constructions.

The key to this inner freedom lies in realizing "emptiness," the essential nonsolidity of things, including thoughts themselves, and their dependence on countless other forces.[6] What we take as solidity—objects, thoughts, ourselves—are *seeming* entities that arise as part of a larger web of connection. Every object breaks down into its parts. Every thought arises and dissolves. Our very sense of self is built from dozens of separate mental processes.

We can think of emptiness in terms of phase transitions in physics. Water, for example, can take several forms: liquid, steam, ice, clouds, or snow.[7] All these are H_2O, but those two hydrogen and single oxygen atoms themselves break down into subatomic particles, which are in essence empty—and dependent on other forces.

Wise Reflection

At the heart of Buddhist teachings there is the uncompromising truth of suffering—of our everyday unsatisfactory reality, that things don't work out as we wish—and the possibility of liberation from it. Seen in this way, to the degree we can lessen the hold of our dysfunctional mode habits, one source of this suffering can be lessened.

The Buddha's path to lessening suffering includes four kinds of mindfulness. First is mindfulness of the body. Second, mindfulness of feelings (which we explored in the last chapter) brings an alert, impartial awareness to our emotions. And third, mindfulness of thoughts does the same for the flow of thought.

The fourth variety of mindfulness brings attention to the way things are, the natural laws that govern our experiences. So, for example, when you bring mindfulness to an emotion and know that it has this or that quality, you are meditating on feelings. But when you know *this is a feeling and nothing else,* you are being mindful of how our experiences operate.

Contrast that clear perception with how we see the world through the filters of our usual modes, with the thick screen of our projections, hopes, and fears, and all the other mode distortions. A wise reflection lets us distinguish between the two ways of seeing. One lets us see how our modes distort perception; in the other, we see only through that distorted lens.

Being mindful itself clears our windows of perception and opens the way to change. So with the tendency to binge on junk food, for instance; realizing when we are in the grip of craving gives us the opportunity to replace that habit with a wiser one—moderation and restraint.

Or take the other route: ill will, resentment, and anger, which can also manifest as anxiety, tension, frustration, or impatience. With wise reflection we can recognize both when these flavor our perceptions and when they are absent. Then, when they begin to arise in our minds, if we can bring them within the beam of mindfulness, they will dissipate.

Then there are deeper truths, like impermanence. When we reflect wisely on this, our relationship to anger—or craving—changes. Unlike our modes, which intensify such feelings, remembering impermanence can undermine such feelings.

Wise reflection also includes focusing on the antidotes to mode reactions. So if, for example, a mode makes us irritated or angry, remembering lovingkindness practice can alter the emotional equation.

The basic principle is that when wisdom arises in the mind, there cannot be delusion. Wisdom, in the sense of seeing clearly, lets us know when our consciousness is colored by aversion, by craving, or bewildered by the distortions of a mode.

Signs of Progress

The only thing permanent about our behavior, a saying holds, is the belief that it is so. As our negative modes begin to weaken their grip, we can sense how we feel freer in aspects of our lives—especially the places where those modes held us back most strongly. We find ourselves spontaneously responding more freely at times when before we had felt imprisoned.

Our emotional habits don't change overnight, but with continued mode work they gradually lose their power. We can get a rough gauge of our progress by asking ourselves:

Am I less reactive and more resilient? More understanding, kind, clear, tolerant?

Do I spend less time caught up in preoccupations and self-concerns and have more time available for others?

Do I have fewer conflicts or less tension with the people in my life?

When a negative mode stirs, do I recognize it in myself? In other people?

Am I less defined by my limiting modes and more understanding of others when they are in the grip of their modes?

Is the secure mode my reference point more frequently than my insecure modes?

Is my attention more free for creative, meaningful, and compassionate pursuits?

Practice Guidance

"I wish I could show, when you are lonely or in darkness, the astonishing light of your own Being."

—Hafiz

Modes manifest differently from person to person and time to time. There is no formulaic method for mode work in any and all situations. But there are basic practices and perspectives we can become familiar with and apply as needed. I'll step out of the way so you can freely try them out, even play with them to make them your own.

Outer Space, Inner Space

Tsoknyi Rinpoche was "playing meditation" with a seven-year-old who showed some interest. He asked her what she did when she meditated. She replied, "I close my eyes and sit still."

He suggested to her, "Look out into space—that's outer space . . . now look inside, at inner space."

She repeated to herself "outer space, inner space . . . outer space, inner space" as she focused intently.

Tsoknyi's father, Tulku Urgyen Rinpoche, tells about how when he was a little boy in Tibet he used to go off to a quiet place near a mountain and pretend to meditate. He later spent years in retreat and eventually became a highly accomplished meditation teacher. Tulku Urgyen referred to inner space as "looking toward the perceiving mind." He once said, in effect, that people think it's the object of attention that's important, but it's actually looking toward where objects are reflected in perception that's important.

Look at a flower. Then look at the mind that perceives the flower.

Ordinarily our minds automatically generate a stream of thoughts. Starting with any single thought, there will be a chain of association to other thoughts, memories long forgotten, captivating fantasies, and reactions. We can look at these thoughts. Or we can look at what thinks.

At any point we can make use of our thoughts as practice opportunities by letting there be space around a thought—suspending judgments of the aversive mind and preferences of the wanting mind, in an open awareness.

In Japanese flower arrangement, the space around the flowers is one of the elements you consider to be part of the overall arrangement. Open space lets us see the graceful curve of a branch, the delicate shoots of greens that frame the petals in bloom.

Space in the mind allows room, which lets us see things more clearly. Then, as thoughts and feelings express themselves, they bloom within the open space of our minds, where we can notice things we hadn't before, as the perceiving mind is gently embraced by spaciousness.

Subtle Mode Primes

During a tornado watch, a circulating cloud indicates a weather pattern that warns a tornado may be forming. That's an apt visual metaphor for what goes on in our minds as an inner storm gathers momentum.

As our mindful radar sharpens through practice, a circulating cloud within gives us time to consider de-escalating. We can close up the windows and wait out the storm until the tumult passes. We might turn to the shelter of a calming or compassion practice or challenge our stormy assumptions. The key lies in applying mindfulness as an early warning radar, spotting negative modes while they are still on the distant horizon.

You can practice integrating awareness with subtle mode primes by paying close attention to any intentions that arise during a meditation session—the general capacity to attune to subtleties will strengthen. In mode work, this could mean catching a triggering thought or noticing a feeling being primed that can activate the entire mode.

Observe, as your awareness follows along, the flow of each changing moment of experience as you sit in meditation. At some point a sense of physical discomfort will appear. Bring a carefully observing attention to bear.

You might find, for instance, that you feel like adjusting your position to one that's more comfortable. But instead, decide to stay in the original position. Let your awareness be with the sensations of discomfort and your intentions to shift position. If possible, try to simply note them, no matter how urgent the impulse to move becomes. Your aim is to stay mindful of the urge and discomfort, but not let it lead to an action.

This is a microcosm of a mode prime. If you can simply bring mindful presence to the trigger, but stop the usual mode reactions, you weaken the mode's chain of cause-and-effect.

From Bewilderment to Wakefulness

It's important not to rush from one mode reaction to another, leaving a trail of distorted assumptions and even more reactions. Where does awareness go when that happens?

Say during a simple act like putting the water for tea on to boil, you start thinking about other things that need to be done. It's more challenging to sustain awareness when there are many things to do and think about, especially if they need to be done soon.

But with mindfulness you might become aware of your mind speeding up and starting to lose full clarity. The overseer doesn't fall for the mode's cover story. It distinguishes the difference between a mode's lens, assumptions, and reactions as well as what's true. It gives more power to a mindful awareness than to a mode's habits.

When we bring mindful habit change to a mode reaction, we can reframe the situation in a more helpful way. First, notice that familiar mode reactions have started. If you catch them before the momentum of the mode habits take over, you will be better able to reconnect with the mindful overseer.

A mindful reminder lets us calm down and clear our minds so we can assess things more clearly; it enhances awareness of what we think, say, and do. This allows us to counter the thoughts, feelings, and reactions typical of the mode itself.

If the effects of a mode reaction linger, we can reconnect with awareness, pausing and reflecting on how to direct our attention and our efforts. We might, for example, decide to take short practice breaks to come back to neutral.

Then we can ask, *Is there anything that might be learned from this to lessen the momentum of these mode habits?* We can pause and reflect on how to make use of this experience as an opportunity to recognize how the mind works when a mode asserts itself.

Common sense guides us though our lives day to day. That's how we make reasonable decisions and intelligent choices based on gathered facts. But mindful habit change takes a quality of what might be called "uncommon sense," a willingness to think outside the box and venture into unknown territories of the mind and heart—unfamiliar spaces.

This means a willingness to change a habit of bewilderment into one of wakefulness.

Staying Awake

When it comes to sustaining mindfulness as your negative modes activate, this strength gives you added options. You can more readily engage the mindful overseer to check what's happening, tune in to see what's needed, and sustain mindful presence.

When strong feelings arise, you can tune in to how you are experiencing them. Turbulent emotions are signals that a mode has been activated. But if the noise and agitation overtake your mindfulness, come back to the natural flow of your breath for a while to anchor your attention. Then open your awareness and be present to the feelings once again. As they play out their cycles, maintain a mindful awareness until they naturally subside. Tune in to see if there are subtle echoes. These can come in the form of lingering liking, disliking, or disinterest.

This can be another mindful habit change opportunity: to take any experience and turn it into a mindful attunement and discernment moment through detecting whether you are feeling liking, disliking, or indifference.

Then you can respond with continued awareness rather than having these subtle reactions prime preferences or judgments, and rather than giving power to them by letting them propel you to action.

The habit of mindfulness can extend more and more into your day if you make the effort. The capacity to be mindful builds in strength the more you practice it, until ideally it becomes a habit itself.

Exploring Dissatisfaction

You can create an opportunity to strengthen mode mindfulness during a meditation by telling yourself not to move, no matter what. Invariably after a while some pain will arise in your knee or back.

The discomfort of not moving opens another possibility: you can investigate the subtle sense of the "uneasiness of existence," all those subtle habits we have in life of moving away from discomforts and adjusting things to minimize dissatisfaction.

By being mindful of these mental currents, you attune to the dissatisfied mind. That aspect of mind very often holds court and calls the shots whenever something doesn't feel pleasing.

If you just need to adjust your position to be more comfortable, by all means, go for it. But if you'd like to play with exploring the dissatisfied mind, don't move. This sends the message that the dissatisfied mind won't always get its way. You're going to choose mindful awareness rather than just playing out this pervasive habit again.

The dissatisfied mind doesn't like this, but it will respect your decision if you are firm. This subtle inner habit change transfers power from the faulty leadership of the dissatisfied mode of being to the executive overseer of mindfulness.

Tolerating discomforts and dissatisfaction means not trying to fix or control things, but just being present with feelings as they are, with a steady attention that can look into the feelings without being reactive.

If you sustain attention you may see what is needed at a more subtle level—sometimes the cognitive, sometimes the emotional. Or there may be a need to let go of an assumption or expectation, or to grieve a loss.

Be patient while going through the unpleasantness. Be especially attentive to aversion in your mind. Look into the nature of unpleasant feelings and aversive thoughts.

One point of this practice is to mindfully change the habit of avoiding unpleasant experiences by noticing the times your mind starts to turn away and instead stay with the unpleasantness. Any time you give more power to your awareness than to a negative mode, it's a small step toward transformation.

Clarifying Anger

The first question about angry feelings is, are they appropriate? If you or someone you care about is being treated unjustly, it's natural to become indignant.

But more often than not, much of anger is a projection and an exaggeration; our core beliefs amplify this, giving the event an added symbolic meaning, which fuels the outrage.

Applying a discerning awareness—if we can at that moment; or if not, then in retrospect after cooling down—helps determine if the anger is an overreaction.

Inappropriate anger only harms our relationships and us. We need to put anger in its proper place by acknowledging how we are interpreting events and discovering if there are distortions at work.

Remind yourself of a mode's network of beliefs and the symbolic meanings that lens may project on what's going on. Empathize with your reactivity and its roots as you inquire into your feelings. Empathizing with symbolic meanings of the mode's history doesn't mean condoning them but rather understanding them.

Ask: *How am I justifying this anger? If anger is mostly a projection and exaggeration of what's really happening, how can I recognize how I am distorting this situation?*

Be present with a nonreactive equanimity, if you can, while you're feeling angry. Without getting caught up and distracted by what you are angry about—or the person you are angry with—just be present in your awareness with the feelings.

Try to sustain a close observation of your experience, attuning to your agitated mind and unsettled feelings or sensations—perhaps there is a pressure in your chest or tension in your jaw—all without reacting and fueling the anger more.

When angry thoughts come to mind, let them pass. See if you can challenge them and identify possible distortions or projections. Question their assumptions.

Perhaps underlying your anger is a reasonable need or purpose. Are there other avenues you might pursue, if you can drop the anger?

How can you use these feelings as a practice opportunity, to allow a sense of clarity?

Allow any messages or insights to come into clearer focus. Perhaps there's some assertive action you can take to remedy the situation more skillfully and respond with awareness rather than just reacting angrily.

Awareness awakens the possibility for insights to be revealed through the heart of an emotion, if we develop a healthy relationship to our feelings.

PART

Tending to the Whispers of the World

16

Two of Me, Two of You

"There are two of me and two of you," a song by Jackson Browne laments. "Two who have betrayed love, and two who have been true."[1]

I take this as an inspiring metaphor for how we all switch into and out of modes from time to time, particularly in our most intimate relationships—with partners, children, siblings, and close friends. Passing modes make someone a "different person" for the time being, a change we can recognize more readily in those we know the best. Mode triggers lurk in our closest relationships, making love an emotional minefield.

We feel most in love and connected with our partners while we're both in positive modes. But our partners can seem befuddling, frustrating, or downright unappealing when they flip into a negative mode (and, of course, we can seem the same to them).

Whether or not we enjoy being with our partners depends to a great extent on the modes they are typically in, as well as the modes that are triggered in us when we're with them.

Marital researcher John Gottman has found that happily married couples have at least five positive experiences together for every upsetting one. The couples who steadily fall below that ratio are more likely to split up in coming years unless they seek help.[2]

This positive-to-negative measure may offer a reading of how often one or the other partner gets in a toxic mode and how often in a positive one. It also suggests that couples who prime positive modes in each other are more likely to last.

Since our modes are learned habits, relationships offer unique opportunities. A client had never heard of the vulnerable side of the anxious mode, but when he learned about it he realized that this captured his wife's behavior at times. They loved each other, and were usually very affectionate, but sometimes things were very different.

His work often required him to take trips out of town, and he was always puzzled that when he would return from a business trip his wife would be emotionally distant for days. She would take a long time to warm up to her usual affectionate self once again.

As he realized how modes operate he was able to see how his trips were a mode trigger for her. She reacted to what she perceived as abandonment by going into a defensive form of the anxious mode, where her feelings of being let down by his absence made her so unnervingly cool when he returned.

Once my client understood this, he no longer took as personally her iciness on his return. He saw there were two parts of her—one the distant wife, while in her hurt mode, the other warm and connected in secure mode.

So with that in mind, the next time he returned from a lengthy business trip he tried to be patient. He waited until her usual loving affection had returned, when the warmth of his presence had melted her frozen heart, allowing her to relax and reconnect. Simply his being there, he came to realize, was all she needed to return to secure mode.

At a deep emotional level a loving partner represents a significant prime for our secure base: the person in our lives we can feel truly safe with. In an ideal relationship, each partner provides a steady secure base for the other, each giving and receiving emotional support and protection. Partners enjoy closeness and connection but can just as easily be independent without loss of a sense of intimacy.

But given how life brings us unexpected mode triggers, that ideal may be hard to maintain. In those times when you or your mate are not able to be a secure base for one another, it can be clarifying to remember that there are two of both of you—and that you can be an active agent in helping prime your partner's (or your own) better half.

In any of our relationships, our judgments (*he's so uptight, she's so defensive*) really refer only to "one" of that person, a mode that person inhabits from time to time rather than who he or she is all the time. We can too readily assume our negative interactions with someone define the whole person or the relationship.

If we step back and give it more room we might see that temporary circumstances have triggered a mode, and that it's the mode we are having the problem with, not the person. So in the heat of an argument one partner might stop and lightly suggest, "Let's attack the mode, not each other."

While it's easier to see mode shifts in someone else, a reframe can show us our own two sides. "The way I was with my ex-husband, and all the men in my life, was that whoever was next to me at a specific moment was credited or blamed with whatever was going on with me," admits Elizabeth Gilbert, author of *Eat, Pray, Love*.[3] "If I was happy, then that person was terrific. If I was unhappy, that person was a jerk."

Now happily married, Gilbert adds, "I won't do that to anybody anymore, much less the person I love and care about the most."

When Modes Fall in Love

Some attractions can be seen through a mode lens. The cool and calm of the avoidant mode, for example, might seem an enticing relief to someone who spends way too much time in anxious rumination. So the anxious partner benefits from learning how to be self-contained, less reactive, and more content in herself—all positive attributes of the avoidant mode.

Likewise, the lively sensitivity of the person whose feelings run high might be intriguingly spontaneous to someone whose favored mode tends toward the overly inhibited avoidant. Such a person benefits from gaining comfort in a more adaptive mode, where she has the chance to learn how to connect, to be more emotionally attuned, and to be comfortable with strong emotions—positive aspects of the anxious mode.

An upside of any special chemistry between people prone to the avoidant and the anxious modes is the potential for reparative learning each might get from the other. The downside is that they will trigger each other's worst modes, making them both a little crazy!

The averages seem to favor reparative learning rather than disaster: research tracking couples over decades finds that partners in long-term marriages tend to become more like each other in many ways.[4] With luck, one of the ways they become similar would be to acquire some of the mode strengths each offers the other.

We are not, of course, doomed to whatever troubling modes we happen to favor. The positives in our relationships can themselves be shifters into the secure mode.

A high-strung friend married to a mellow guy once complained, "My sisters and I come from a family with a strong strain of neurotic melodrama. But there's something I can take for that."

"What do you take?" I asked her.

"My husband."

Couple Modes

Modes are particularly tricky when it comes to our partners, in part because that connection in adulthood tends to echo the emotional dynamic—and dominant modes—of our parental relationships in childhood.[5] If we had a parent who, for instance, we adapted to by taking refuge in the anxious or the avoidant mode, we are likely to go there with our romantic partners, particularly when we feel that the relationship itself is somehow threatened.

One relationship wrinkle caused by the "two parts effect" can be seen when a romantic partner (or close friend or boss) is mercurial, shifting suddenly and unpredictably from a good mode to a toxic one. The partners of such erratic mode-flippers can't trust which mode they will be getting, even in the most seemingly innocent encounters; they feel constant wariness along with a low-level apprehension.

When one hundred men and women had blood pressure readings taken while they were having enjoyable interactions with family or friends, their blood pressures were found to have fallen. When they had troublesome interactions, their blood pressures rose (as might be expected). But if the person they were with was unpredictable—sometimes pleasant, sometimes explosive—that produced the highest rise in blood pressure.[6] When we are fearful that the person we are with might slip into a toxic mode at any minute, the relationship exacts an emotional toll.

When an unpredictable person holds power over us, like a boss, constant wariness can create levels of biological stress where the brain continually pumps hormones that prepare us for a threat, which can also affect our health when sustained for a long time.[7]

Alicia complained of chronic difficulties with her closest female friends. Yet again she found herself caught in worried rumination about a longtime woman friend who had repeated the same pattern over the years. The friend would be "there," very available and warm, then suddenly shift into quite a different mode, where she was indifferent and dismissive, which destabilized Alicia's secure base.

Alicia was seeing how such friends were keeping her own dysfunctional patterns active. She realized that the hope for change and her attachment to connection at any cost begged for a closer examination. *Connected to what?* she asked herself. Her answer: to the continued painful pattern of interaction that hooks us with a promise of changing to, maybe, something different.

The operative term is "maybe," because in the not knowing and the compelling mystery behind how the story will end, we are often left dangling with our hopes and fears. But instead of remaining a pawn in this mode scenario, we can take some control by changing the pattern. If you feel challenged by a trait in someone whom you feel close to, the closeness of your connection—and your sharing a secure mode together—may override the way you're impacted by that annoying trait and minimize its importance. But if your connection is diminished, that challenge may become more predominant.

In romantic life our initial attractions inevitably lead us to idealize the ones we're attracted to and downplay any negatives. There still can be love, but that magical first phase of a relationship inevitably fades as the full reality of our partners' modes begins to sink in. That's the predicament summed up by the anonymous wit who observed, "Marriage is a romance in which the hero dies in the first chapter."

That's when habit change comes in. Can you become mindful of how your partner triggers your negative reaction? Can you both agree to do mode checks from time to time, to monitor your modes together? Can you find ways to alter the habitual mode dance and build a more positive pattern?

Sometimes when we're caught seeing our partners through a negative mode lens, we can benefit from a gentle reminder that we can have two sides. A friend tells me that when she and her husband were first getting together in a long-distance relationship, they wrote passionate love letters to each other. "At times when I hate him, I take out those letters to remember why I love him."

She was intentionally changing how she reacted to her mode triggers. To the extent that we are trapped in maladaptive modes, we need to learn how to handle the ways those modes surface in our relationships. It's part of love's territory.

Together in a Better Mode

In horse whispering you learn that as you approach a horse you should pause and look away for a while rather than advance steadily. Then you can continue on to join up. Horses approach and retreat during their engagements, respecting their need for "bubble time"—space to be on their own.

It's a pattern of advance and retreat that connects in all kinds of relationships. As a toddler gains self-confidence he ventures away from his mother for a while, then returns to the safety of hugging her before moving away again. A couple flirting will make eye contact, then look away before their eyes meet again.

"The more you can be separate, the more you can be together," as one family therapist put it.[8] Every couple negotiates a balance between connection and autonomy, knowing when to approach and when to give space, together and apart.

So long as we stay connected, our advance is not overbearing, and our retreat is not abandonment, but rather respecting emotional needs.

We each differ in these needs, and they are constantly in flux. So finding this balance takes attunement and spaciousness, not a formulaic solution. The challenge is to stay connected at the core in a secure base amidst changing circumstances and fluctuating modes.

Partners need to find the right balance for their relationship.

Isabel told me her husband, Julian, seemed happy spending hours alone, glued to his computer screen, "always wrapped up in a shell," as she put it. Julian was prone to what Jeffrey Young calls the "detached protector" mode, an extreme of the avoidant stance. This continual disconnection left Isabel feeling lonely and ignored, which was doubly painful for her because it echoed her childhood experience of adapting to a workaholic father and a self-absorbed mother who had ignored her much of the time.

Frequently Isabel would reach a point where she couldn't stand Julian's emotional distance and would launch into an attack. She relied on the favored strategy of the anxious mode, turning up the volume on her complaints, assuming it was the only way to get his attention. But often she overdid it. She'd get so angry she would shout—attacking him, not addressing his behavior. That inevitably drove him into a quiet panic. He'd go absolutely silent, wanting desperately to head for the nearest exit.

I advised Isabel to let him know about what she needed from him in a way that would allow him to give it and to stay focused on what he did rather than making a personal attack. She needed to be firm and assertive about her feelings rather than losing her temper and threatening him.

"When I see you at your computer," she said to him, "at first I think, 'Oh, this is a good time for me to get online too.' But then when you spend a lot of time with your laptop, I begin to feel that you'd prefer to be with your computer more than with me. Can we figure out a way to get both our needs met—quality time together as well as the time we need to work at our computers?

"Otherwise," she added wryly, "it seems like world peace is at stake in this very household."

With Isabel replacing their old confront–avoid pattern with this more constructive, witty one, for once Julian didn't feel the need to run away. He was intrigued by how skillfully Isabel had couched their problem. Julian closed his laptop and gave her his full attention, feeling receptive and engaged rather than attacked.

As Isabel explained the modes to him, Julian recognized that he fit the avoidant in most every detail. Isabel said, "I'm aware of that, and it helps me not to take what look like rejections so personally. I know you need more space and I love my solitude too, especially when I'm not feeling alienated from you."

She went on to explain what her anxious concerns around connection are like. "I even get bent out of shape if our cat seems to be ignoring me!"

At that, Julian laughed at her ability to see the humor in it all, which signaled an easing of the anxiety that underlay his avoidant mode.

As they talked it through, Julian and Isabel realized that when their anxious and avoidant modes evaporated, they enjoyed each other's company and their engaging conversations. They saw how these modes were obstacles to their loving, playful connectedness.

Julian and Isabel were intrigued by the mode-change principles. As an experiment, Julian decided to build a timer into his computer, one that reminded him to take a break and spend some time with his wife. He liked the challenge of integrating these awareness tools into his routine to change his mode habits—something like a mental app. Besides, his wife was cuddlier and softer than his laptop.

When Isabel wasn't in the midst of her anxious, clingy mode, she could be emotionally astute and articulate about a realm that Julian found a bit perplexing. Like many prone to the avoidant mode, Julian was intrigued by her ability at these times to investigate emotions in a balanced way rather than his usual run from their intensity. He actually enjoyed learning this from her and even felt relieved that they could be together in this relaxed way.

Isabel, too, made an intentional mode shift: she looked for ways to be kind to him. She knew that Julian was a bit of an information junkie and that he enjoyed reading interesting articles people posted on sources like Facebook. She was touched by his using a timer as a reminder to get together with her. So she changed this into a mutually rewarding routine: when they got together, they would go for a walk in the park, where she would ask him to share with her what he had been reading.

They were learning how to appreciate each other's positive mode strengths and changing their mode habits. Instead of their old mode dance—with her anxious clinging triggering his avoidance—they were forming a new habit of enriching engagement.

Genuinely compassionate efforts like this can help rebuild trust and goodwill in any relationship. When it comes to releasing the grip of a maladaptive mode in a relationship, we can help each other. We can avoid triggering such modes in the other person. Or, as Thich Nhat Hanh says, "You must love in such a way that the person you love feels free."[9]

It's helpful to remember in any relationship that we all perceive and interpret things so differently from one another. Remembering this can lower the risk of conflicts in our relationships.

Mode Lock

The movie *The Story of the Weeping Camel*, set in Mongolia, tells the story of a baby camel that was rejected by his mother immediately after his difficult birth. The mother would not let him nurse, so the baby was bottle-fed by the nomads who tended their herd, but he had no secure sense of connection with his mother. He was always off grazing by himself, not with her.

Out of concern for the baby, the nomads asked a musician, a kind of healer, to come help. The musical shaman attuned his stringed instrument to the mother camel's natural sounds and played them back to her in the same frequency and tone. As the mother camel felt heard, she experienced a bonding resonance, and a tear fell from her eye. After that, she began to nurse her baby.

We can use an analogy of resonance to reconnect. In resonance, one object vibrates at the same frequency as another. This entrainment was observed long ago with the pendulums of Swiss clocks left in the same room together. When there are two pendulums swinging next to each other at different rates of speed, the faster pendulum will shift its rate to synchronize with the one that is moving in slower, wider arcs. Soon they will be entrained, moving at the identical, slower rhythm, in a state physicists call "mode locking."

The bodywork method craniosacral therapy applies this principle. If the craniosacral therapist can sustain an open, slower rhythm in his mode of being, this helps a client synchronize—slow down to a more receptive pace conducive to healing. The same holds true in our relationships. When one person can sustain a steady and open presence, that stable mode can be transmitted to the other.

Modes, like moods, can be contagious. Someone who is connected to her secure mode can be a soothing influence on us, simply through her mere presence—just as a parent's touch calms an upset child or a loved one at a hospital bedside offers a reassuring presence to a patient.

If we are stable in our own positive mode, we can be senders of good feeling, not just passive absorbers of other people's bad moods. This can give us a larger range of choices in how we take in their states, what feelings are evoked in us, and how we respond.

This protects us from being at the whim of whatever feelings arise when we happen to encounter what might have been a mode trigger. We can keep our cool and, with a mode lock, help a friend who is melting down be calm rather than get caught up in that person's distress. For those of us in the helping professions, this inner capacity lets us understand and empathize with our clients, but not be overwhelmed by their distress. Instead we can remain stable in our own secure mode and from there do whatever will help them calm down and look into the source of their troubles.

Mode locking comes up in our everyday connections. Two women who live in distant cities visited each other after not having gotten together for a long time. They were old friends, almost like sisters, who mostly had gotten along well. But—like sisters—they would have their usual disagreements, mostly about the minor details of arranging their visits.

During this visit one of the friends had just been on a long, nourishing meditation retreat and was feeling replete. When their usual this-could-get-complicated arrangements were being made, the predictable issues that could have turned into a spat seemed insignificant to the one who had come from the retreat. She just felt so happy to be seeing her good friend that when it came time to iron out the arrangements, she said in an easygoing tone, "Whatever works is fine with me. What's important is being together."

Over the course of their time together this mode prevailed, changing the way they related to each other. To be sure, little disagreements arose. But what would usually have triggered a minor battle seemed like small stuff compared to the joy they felt just being together.

With entrainment, by resonating with a secure mode ourselves, we can invite those we love to join us.

We Are Not Our Modes

Sometimes mode changes are like discarding shoes that no longer fit. This happens naturally in the course of life, when we get to a point where an emotional habit no longer feels like "me." We've changed from the inside and now see things—and ourselves—differently. But from the outside people in our lives may still assume we're the same and may still treat us like that old "me."

A mother had raised two daughters, who had gone off to college a few years before, leaving her with open time. She had been utterly devoted to raising her daughters, sometimes in the overly compliant, prey-like mode. But now, at last, she loved having time to herself to read, meditate, and work on her own projects.

When her daughters came home from college, they brought their old expectations along with them. Though they were now grown, they treated her in the same old way, expecting her to drop everything to tend to their needs. So often you hear newly independent young adults say how they no longer fit in the same small world of their families. But here was a mother who said, "I sometimes wonder if they see that I've changed. I don't feel like that old me anymore. It's a bit of a struggle to get to know each other in a new way when your old habits of relating are so strong."

One tricky part of this work is the secondary gain that negative modes can give, the subtle dynamics that keep them in place despite their drawbacks. For example, if we are nurturing caretakers sacrificing our own needs to make others happy, when we start to include ourselves among those whose needs require attention, some people in our lives may not like the change.

Or if we are perfectionists whose ultra-hard work has made us successful, but at the expense of other needs, like our health or our relationships, we may find an inner resistance to rebalancing our time and energy. We may find it hard to simply accept ourselves (or others) as we are rather than being critical. It might seem to threaten the identity built on success or perfection.

Everyone has positive and negative modes, and in their negative ones they are more likely to act in ways that upset us. If we were to encounter the same person while he (or we) was in a positive mode, we might like him. So if there's a building annoyance from the way he is treating us, instead of a knee-jerk reaction (*how inconsiderate!*) we can take a moment to remember that it's not the person but his action that's so irritating.

We're all a mix. The Dalai Lama offers this advice: "If you see a person as 100 percent negative, when you check more closely you will see it isn't so. Likewise, a person is not 100 percent attractive or unattractive, pleasant or unpleasant. It's projected mentally—by you. Through an investigative inquiry you can reduce your attachment or aversion toward the person."

This sort of reappraisal lets us live more wisely in our relationships. Awareness can give us the inner space that lets us replace an automatic negative perception of someone with a more charitable, constructive outlook.

Two old friends were locking horns. One said, "This is not about you!" The other fired back, "It's not about you either!" They both cracked up.

The key point is this: you are not your modes, and neither am I. This insight can help us recognize the difference between our mental model of a person—which assumes he or she inhabits a given set of familiar modes—and the actual person. Our assumptions about another person's modes subtly determine what we expect of and how we treat her, which can, ironically, cue that person into a habitual undesirable mode.

In any of our relationships where one negative mode chronically triggers another one, that toxic pattern starts to loosen whenever one of us can instead make changes that are likely to lead to more positive connections.

Take, for example, parents and children. So much of what we believe about ourselves may come from what our parents believe about themselves, which in turn may have come from their parents. Each generation inadvertently programs the next generation.

Seeing that people are not their modes allows them the potential to inhabit a different, better set. By changing our own expectations of others, we let ourselves be with them in new ways that can, in turn, allow the emergence of a different range in our relationships.

My friend KD had a clarifying insight while caring for his mother as she was dying of lung cancer.[10] For the final few weeks of her life, he moved into the guest room of her apartment and took care of her day and night.

"I realized that I had waited my whole life for a chance to do something for her," KD said. "She had always kept me at a distance but, in her illness, she was too weak to push me away. She was sick and I was there, and she was letting herself enjoy our being together.

"We got very close at the end. She told me I was the best nurse she ever had. When she was shifted to the hospital, I would sit with her late into the night, being with her in a deep, sweet flow of love.

"But she was still my same old mom! One time she asked for a drink of water and, as I leaned over her, I knocked against the tray on her lap. She snapped at me ferociously and I recoiled as if I had been hit in the stomach. As I stood back up, I had a sobering realization: from the time I was a small child, even before I could remember, she must have been doing this to me in just the same way.

"The fact that I couldn't remember it was a shock to my system. I realized that I couldn't remember it because I had taken shape as a person in reaction to the way she had treated me. My emotional shape was so deeply formed that it couldn't see *itself*! Nonetheless it still had to be functioning under the surface. I was stunned and amazed.

"During her last days, I felt as if we truly met and both loved and forgave each other fully. When she finally passed, I had another extraordinary realization. I saw that my mother had been like a huge electromagnet and I was held in an emotional shape in much the same way as iron filings take the shape of the magnetic field around a magnet.

"With her death, the plug on that magnet was pulled and my iron filings relaxed into a more natural shape for me. The force that had been pulling and shaping me since I was a baby was gone; I felt as if I had been given much more comfortable clothes to wear, clothes that fit me as I am. I was free to be myself and to breathe in a new and different way. We had freed each other with love and were no longer chained together by our emotional programming."

CHAPTER 17

Joined at the Heart

Bob asked if I wanted to teach Sandhi how to pick me up for a ride from where I was sitting atop a fence on the edge of the riding ring. So I saddled her up, asked her to "Stay," and left her standing on the opposite side of the ring from where I hopped onto the fence.

Usually she would just follow me or wander off herself, but we were already engaged in a conversation and she was waiting for what would come next.

Then I faced her and called her over to me. Once she started walking over, I turned slightly, signaling to her that she understood what I was asking—communicating with her in a language she understands.

After she positioned her body, parking herself directly underneath me, she stood perfectly still. I gently slid onto the saddle she was offering me, and stroked her mane, telling her what a good girl she was.

"She doesn't want to be anywhere else in the world right now," Bob said. This was not a small thing, since it was during her feeding time!

I, too, could feel an invisible bond between us, a sense of connection: riding on her back while joined at the heart. I became part of her herd, a being with whom she feels safe and relaxed.

The phrase used in horse whispering for this special quality of human-and-horse connection, as I've mentioned earlier in the book, is "joining up." In-the-moment, attuned connection is the essence of joining up. It is also a key part of mind whispering as it applies to our relationships. When we pause and pay full attention to the other person, we join up, connected and attuned to one another. When two people resonate on the same wavelength, neuroscientists say, in a sense their two brains are coordinating into a single complex system.

When we do so fully, we experience what I call the physics of connection. With that kind of rapport, we can stay joined up in any relationship, even when we are occupied with other things or are far away.

Just as the first step in mind whispering is a kind of joining up—attuning to our own minds—here we apply the same full attention to the other person. Daniel Siegel, a UCLA psychiatrist, describes how the same brain circuitry we use to empathize with another person gets applied to tuning in to ourselves when we become mindful.[1]

This is the I–You relationship, in which we fully orient to one another and feel seen and cared for.[2] Joining up was portrayed vividly in the movie *Avatar*, when the native Na'vi people would connect the tendrils at the tips of their tails with the tendrils of the giant birds and horses they rode—or with each other.

The I–You takes many forms. "I always like it when my husband tells me he loves me," a friend says. "But I *really* feel seen and cared about when he orders for me and knows that I like my huevos rancheros done over easy, no sour cream, and salsa on the side."

My grandmother as far back as I can remember was always a health food and exercise enthusiast. When I'd visit her as a kid I would get discs of sweet acerola vitamin C tablets instead of candy, followed by long beach walks.

Some years ago I got a call saying that my grandmother was in intensive care after a massive stroke. Stunned, I immediately left for the hospital. When I got to her side, I was shocked by how frail she seemed—as if her life were drawing to a close.

I guess because she had always been interested in healing remedies and health care, I started telling her about some new approaches being used to help people with strokes regain much of their function through recent advances in rehab. In my denial of the harsh truth that my grandmother seemed to be fading away, I was searching for something that I could do to help prevent this and help her bounce back to life. She had always been so resilient, even in her later years.

Lost in my meanderings, it took me a while to notice she was just watching me with an almost vacant gaze. When I realized that she didn't seem to respond at all to what I was saying, my heart sank. And then we dropped into a silent exchange, where only our love felt real.

Connecting with a language of the heart, I said lovingly, "I hope you're peaceful through this whole thing." Suddenly what had been a vacant gaze vanished as she got more focused, her eyes now heartfelt and engaged, she squeezed my hand, seeming to join in the bond of love between us.

By the next day she was gone.

Heart Whispering

There are those moments in life when there are just no words to express what we feel, unknown terrains of the heart where silence feels truer than any words—and times when the principle of love listens and attunes to hear the essence of things, of ourselves, of others. This wise love is where intuitive whisperings of the heart find natural expression.

This mode of connection can sometimes arise spontaneously. It is available to us when we can look past our reactivity—and the other person's—to share the mode where we can join up together.

This collaborative stance sharply contrasts with the way horses have been trained over the centuries: using brute force to get the horse to give up. But in horse whispering you're engaged in a joined-up collaboration.

There are direct similarities between how I tune in to my horse and how I tune in to my own mind. For instance, when I ask Sandhi to move toward a particular position, I first remember the instructions and stay focused on the principles that communicate that intent to her. Then I let go of any assumptions about old habits or relating to her and attune to her mind and spirit—connecting through our hearts while communicating in a language that she understands.

When she seamlessly executes what I am asking, as Bob puts it, "That's her way of saying, 'You speak horse!'"

As she follows the instruction, I acknowledge in "horse" that we understand each other, by turning away or stroking her mane. We communicate free from the constrictions and assumptions set in place by humans' controlling, even predatory tendencies, which so often perplex horses.

This takes a kind of yielding, a receptivity softening our minds and hearts.

One day I asked Bob, "What do people experience in life as joining up?"

"First," he said, "we can look at ourselves. If we enter a relationship in a predatory way, it's not connecting with the other person. If it's only that you want something or need something from them, it's not joining up. If you want too much, demand, control, manipulate—that's being predatory, not being in relationship. That disconnects. We need to empathize and know who they are and give to them, to receive anything. It's giving, not taking."

My good friend Elizabeth and I were talking to a wise guide, Khandro Tseringma, about how children can learn to be more aware when they get into conflicts. "If kids are fighting with their friends, they should reflect on whether it's their own mistake or their friend's," she said. "If it's their own, they should apologize."

If a child tends to be bossy, she added, that child could learn to apply it in a positive way—shift his or her attitude instead to being more forthright and clear. A friend's children went to a school with programs in social and emotional learning, and I asked how they were affected. "It's given them the confidence to be themselves," she said.

Being joined at the heart helps us go beyond a predator-like, controlling mode to see another being without interpretive filters. Attuning to our own true selves and to others—past our assumptions, interpretations, personalities, defenses, and reactivity—lets us connect more with what's genuine.

In that space we might recognize that people's perceptions of the same event can vary greatly—we all interpret things differently. When we work with our own minds and reactions first rather than just having a knee-jerk response, it changes our relationship to these issues.

We can stay connected to the person even while disagreeing with that person's point of view. We can turn to the better angels of our nature, open up to more accepting, kind feelings that help us accept the other person's essence, even when we see things differently. That person is more likely to experience this unconditional regard as a positive emotional message, an invitation to join up.

The Two-Way Street

Breaking a horse by using restraints and sheer force is a bit like the predator-like mode in humans, when one person tries to control another. Such behavior reflects what Joanna Macy, an expert on Buddhism and systems theory, calls "linear causality": we think of A as causing B, B then causing C, and so on.[3] In a relationship, this can lead to an I–It predator-like mode—a one-way street.

By contrast, on a two-way street we have mutual respect and a bit of humility. Instead of my trying to coerce you to change in a certain way, I can invite you to reflect with me by asking a question. Instead of imposing on you the direction I select, we explore together.

On a two-way street (with "mutual causality"), the underlying assumption is that I don't have all the answers, but we can investigate the uncertainty together in conversation. Just as in horse whispering, you can guide but you don't force. You appreciate that the other person's perspective could have value and that you might learn from it. You offer the opportunity for a collaborative partnership.

Resistance and ambivalence from a horse signals that communication has shut down, and you need to change your approach. It's the same with people. Predatory behavior can only work when the other person falls into the passive, prey-like mode.

Reflect for a moment on any relationships in your life where you feel yourself falling into the predator-like mode: aggressive, controlling, dominating. Then imagine letting go of that control. See how it feels. Be aware of any clinging to being "right" or in control in any way. Ask yourself how you might be different with that person, by moving toward joint decisions or empathizing with them and asking, "Does this work for you?"

You can also reflect on relationships in which you feel controlled. In the compliant prey-like mode the controlled person experiences resignation, passivity, fear, or sometimes an angry rebelliousness—all prey reflexes. The controlling predator-like mode, like the compliant prey-like mode, can also be fueled by fear. For the predator-like mode, the fear is of losing control; for the prey-like mode, the fear might be of losing connection.

Of course it takes work to unravel these habitual responses, but awareness and effort can help surface the secure mode. From there you can empathize with the other's way of seeing things and feelings. You can be aware without reactivity or judgment. Perhaps you sense the other person's modes at work.

Connect with your own sincere intention. Communicate clearly, leaving added emotions and reactivity out of what you say. "Allow your words to pass through your heart," as an Indian proverb advises.

From this secure base, you can respond differently: invite the other person to join up, and collaborate in making decisions together.

That's like taking the bit out of a horse's mouth. The other person becomes free to make his or her own choices. It unbridles the creative spirit. Ideally, you join up on a two-way street.

Virtual Connections

A cartoon shows what looks like a twelve-step meeting. Everyone looks at the man standing to testify. "Hi . . . my name is Mike and it's been five months since I stopped a conversation so I could check my Blackberry."

How often do you see people walking down the street or sitting in a café together, and one or both of them are talking on a cell phone or typing away on one? Truth be told, I've done this myself any number of times.

But if we tune in to such a moment with the person we're with, we may well sense the I–It disconnectedness, despite all the artful ways we have to reassure each other: "I don't normally do this, but I've been trying to reach this person for days."

Perhaps it's perfectly fine. But the fact remains that even as we're social networking with our Facebook friends, we might be ignoring the person sitting right across from us. The question is whether distressing modes are being primed that we had not realized. Or whether we're trying to escape the one we're with even while we're trying to connect with someone afar.

Such a social moment presents an opportunity to reflect on what may be going on at the mode level. You might ask if, for instance, there are ways you feel more comfortable with people at a slight distance from you—a clue to the avoidant-mode preference or perhaps to some tension in your relationship with the person you've just turned away from.

Our connections can be thought of in terms of approach and avoidance, advance and retreat, moving toward or giving space. If the connection between us is strong, then we can stay connected even over thousands of miles—or while we take a break to check our e-mail. But if the connection is absent, hours together may not make a difference.

Horses rarely stand head-on, facing one another; for a horse that's an invasive, predatory stance. Personally, I like eye contact (when it feels genuine), but I also like space. There's a difference between space and distance. Is there a point in our ongoing connections when someone becomes an "ex" even when we're still in a relationship? Maybe the problem is with the word "relationship"—relations seem fluid while the "ship" seems static.

Then there's the nature of our virtual connections—Facebook, e-mail, and the like. I enjoy e-mailing and social networking with my friends, not just for the pleasure of staying in touch at a distance, but also because it gives me a chance for more reflective pauses and editing time to communicate with more careful consideration of how I put things.

There's a sense of joy at this ongoing interaction with friends all over the world, from diverse backgrounds and of all ages, and it feels like we're offering each other a service when we post interesting articles, pithy quotes, and share useful information and creative expressions.

Even so, I have some concern that our Internet habits can be a training in distractedness, surfing on to the next thing without much time to absorb or reflect. This might interfere with developing wisdom qualities that need more spacious awareness.

We can be more skillful and mindful of what we say online; in fact, we need to, because the words we send don't have our tones of voice, body language, and facial expressions going along with them to add emotional nuance. Mode mindfulness can be a great help with this. If we take the time for a mindful pause while we write a message online and check whether we're writing from a negative mode, we can lessen the likelihood we'll send a "flame"—a message shaped by distorted perceptions or strong negative emotions, one that we'll regret later.

If we take time to check our own mode-ivation and then tune in to how the other person will receive what we write, our communications can carry the nuances we intend. The Internet is a powerful tool, but it all depends on how we use it: as a force that disconnects or one that moves us toward a web of connectedness.

Joining Up Methods

In a sense, Bob explained to me, the way we interact is a language in itself. I'm having a conversation with Sandhi by how I position my body in relation to hers. It's a spatial language in which our bodies are our voices.

Bob adds, "If I don't have this kind of connection with her, we won't be able to work well together. I won't have a good relationship with her unless I understand what's important to her.

"She's a prey animal. If I look directly at her, turning my eyes, shoulders, and core toward her, she sees this as predatory body language. If I force, control, capture, or relentlessly pursue what I want, it's also predator-like—and clueless. I don't want to trigger her prey-like self-preservation reaction, which can be from mild anxiety to out-and-out panic."

"This relationship is based on trust," I commented.

"Exactly," said Bob. "She knows I'm not going to hurt her. Because she trusts me, it gives me a great responsibility. She's surrendered some of her survival mechanisms.

"It's relationship-building," Bob continued. "You need to communicate in a language she understands, with respect. It's almost a spiritual connection. As we learn their ways as prey animals, we can better empathize and understand them, and establish that trust. Then we learn a language we share together—one where she understands and communicates her ideas to me, and I communicate mine to her.

"I bring shame upon myself if I dominate a horse. Do you know why?" Bob asked.

"Because you're not empathizing?"

"Yes. And because they are so willing to learn and collaborate."

While Bob and I were having this conversation Sandhi, who seemed to have gotten bored, had wandered off to graze nearby. "So if I wanted to ask her to stop eating grass right now in a prey-like way, I would go behind her," Bob explained, as he walked behind Sandhi.

At that, Sandhi stopped grazing and turned around to face and relate to Bob. He stroked her mane for answering him, and then said playfully, "I stroke her mane and I'll take her out to lunch!"

Bob walked off with Sandhi glued to his back in a joined-up connection. "I'm using the language of the herd," Bob told me. "Her nature is to be a willing member of society. Prey animals are gregarious. She knows I understand her and right now she'd rather be with me than eating grass."

Then Bob led Sandhi right back to the grass, where Bob bent down and starting pawing the tall grass with his hands. "Now I'm making believe that I'm eating grass—showing her that I'm like her." Sandhi looked at him delightedly and started to graze too. Bob said, "She's saying, 'You are so cool—you're a vegetarian!'"

Just as Bob follows certain principles of horsemanship in joining up with Sandhi, there are methods we can use to join up with each other. For instance, when we work with our own minds and reactions first rather than just having a knee-jerk reaction to another person, it changes our relationships, so we can stay more connected.

Other principles for joining up include recognizing your own reactive thoughts and what has triggered a negative mode. Internally acknowledge your own vulnerabilities and mode distortions in perception and thinking. Then remember the choice you have, and see if you can find a wiser way of viewing things. You might just decide to let it go.

Recognize when you are coming from a subtly controlling stance and instead turn toward the path of joining up via empathic communication; pause to consider how the other person might be seeing the situation, then act in a way that acknowledges that person's view too.

Perry Wood, a British teacher in horsemanship, tells of a time when two women brought their "problem" gelding to a two-day clinic. The problem was that it took hours to get the horse into his trailer—in fact, three hours just to get him on the road to the clinic.[4]

And when the clinic ended, sure enough, hours after everyone else had left, the horse was *still* not in the trailer, despite every effort the two women made. The horse would go halfway in, then rear, back out, and leap around the loading yard. When Wood came upon the scene, the horse was wide-eyed and sweating, and so were the women.

So Wood approached the horse with his vast repertoire of methods, sense of timing, and determination. Raising his energy to match the horse, Wood tried everything he knew—with no luck. The horse would go halfway in, and then in an equine explosion fly back out.

This went on over and over until Wood paused to ask himself what he was missing. What, he wondered, should he do differently? Then he realized he was probably repeating some version of what everyone else did, using various techniques on the horse without regard to whether or not the horse trusted them in the first place.

"What was missing here," says Wood, "was love, relationship, acceptance, trust, and respect."

So the next time the horse went halfway into the trailer, Wood didn't try to get him to go in all the way. Instead he just stood next to the horse and opened his heart, sending him love. At that moment they joined up.

"Time stood still," recalls Wood. "I felt a lump in my throat."

And the horse's eyes softened as he took a huge breath, his sides heaving with relief. The tension and fear drained from his body. Then, without any further urging, the horse walked peacefully into the trailer.

Wood, tears in his eyes, stroked the horse. And the people watching were crying too.

CHAPTER 18

A Shared Secure Base

A preteen boy from a troubled family started a club at school. He organized groups of kids to fight each other. He called it the Fight Club, and he got lots of attention from other kids.

When the school principal found out about the club, the boy was suspended. He then went to stay with a relative who was a social activist and who was kind to him. The relative took the boy along to help serve food at a soup kitchen and to a march protesting a factory that was dumping toxic chemicals in a poor neighborhood.

The boy eventually returned to school and started another club. But this time it was called the Peace Club. He organized groups of kids to sign up to do good works in the local community. He again got a lot of attention from his peers, but this time for positive, meaningful activities.

What made the difference? When the boy went to live with a kind, supportive relative, that person was able to help him connect with a secure mode. So often when people act in angry, antisocial ways they are being ruled by their negative modes. If they can connect with and nurture a secure mode, there can be a perspective shift that allows them to act in more positive ways.

From the perspective of joining up, whether one-to-one or collectively, we can share our secure mode. Many of the problems society faces are the public face of private realities, with negative modes as hidden drivers. But if people are able to join up with someone into a secure mode, our perspectives shift and those problematic behaviors often seem just what they are: pointless.

And even when we seem most separate, we can still trust in an underlying interconnectedness. This doesn't mean we don't disagree on issues, but that's about the issues; it doesn't impair our essential connection.

These disconnections are often activated by our modes and the ways we see and react when we are influenced by them. So if we don't get along, the problem may be the modes that are driving us, not the person. But if we can join up in a shared secure mode, then we are far more likely to resolve problems between us—whether in a couple or family, between friends or business partners, or in our communities.

From our secure mode, we're better able to give people the space to be heard. Empathy lets us step into the other's shoes to understand that person's feelings and point of view.

There are three kinds of empathy. The first is purely intellectual: we sense how the other person thinks about things and so can take that person's perspective. If cognitive empathy combines with heartlessness, the other is more like an object to be used. But when we take the other person's perspective to heart, we start to join up.

Brain scientists have discovered a type of brain cell, mirror neurons, which may explain some of the magic of the I–You connection. If we were talking, my mirror neurons would act like radar, sensing how you are feeling, your movements, even your intentions, and, mirror-like, they would activate in my brain the same circuitry active in yours. This creates in me an inner sense of what's going on with you.

The "with heart" comes along with emotional empathy, the second variety. This sort of connection means that our brains' social circuits are fully attuned to the other person, so we feel a resonance, sensing immediately how the other person feels. This emotional link keeps us attuned to one another, creating the interpersonal chemistry at the heart of joining up.

The third kind, empathic concern, arises from the first two: we understand how the other person sees things, sense how that person feels about it all—and if he or she is suffering or in need, we spontaneously do whatever we can to help. This adds caring to the mix, making us a secure base for others.

All three empathy types were at play when I was talking with a longtime client, Rosalie, who has had a lifelong tendency to be harshly critical of herself—a perfectionistic, self-judgmental mode she directs occasionally at others but mostly at herself.

I was listening with an open heart to her long list of faults, which she was trying to get me to see in her, almost convincingly. As she spoke I had a mix of feelings: I wanted to make an effort to empathize with her perspective on her litany of faults, and at the same time I felt an increasing sense of my heart warming with the care and compassion I felt for her.

As tears rolled down her cheeks, and after I had listened quietly for a while, I said, "I'm really trying to understand and share this view you have of yourself, but I just don't see you in the negative ways you do." More tears came as I continued. "I've known you for a long time and I know your mode distortions quite well. I just don't think what you are saying is accurate. Who has set up these unattainable standards for you? I can't think of anyone who sees your faults the way you describe them. And you don't mention even a single positive quality out of your many wonderful ones—your generosity, caring, intelligence, and bright spirit."

Perhaps because Rosalie knows how much I care about her and feels a bond of trust with me, in that moment she started to believe me more than what she had been telling herself. She saw that it was true—she hadn't mentioned anything positive—and that maybe her thoughts had been imprisoned by this negative self-perception. I watched her mood shift, almost like a heavy veil lifting from her.

It helped to empathize and listen to her, to express both cognitive and emotional empathy, and to try to understand her perspective while she looked through the veils of her negative projections to her natural qualities—all with a tone of warmth and connectedness.

When we join up in this way, it's important to empathize and understand first, before making a cognitive shift to a rational counter-argument. Otherwise the person can feel her emotions are being dismissed or judged. Particularly when we are dealing with our own or someone else's deeper emotional patterns, creating a sense of a shared secure base creates a safer space within which you might, for example, counter someone's negative beliefs with a sincere, caring heart. A joined-up connection melts inner and outer barriers to a more accurate way of perceiving.

A Gentle Path

"I like to be straightforward in my relationships," a young woman confided, "but it seems to scare men away!"

The guy she was with now, she added, really seemed willing to look at himself. "We have pretty honest conversations. He takes many things to heart and works on them. But after a serious talk we had recently, where I said some things that may have seemed intense, I saw him wince. He said, 'That was hard to hear. Can you find a way to be more gentle with your honesty?'"

Mahatma Gandhi observed, "In a gentle way we can change the world." The same holds true for us. There is much we can learn from the cooperative strengths of prey species (and the prey-like human mode). In a sense this represents the positive end of the spectrum of cooperation— not a compliant surrender but the collaborative spirit, responding to the collective needs of the group. With an implicit understanding of interdependence, we don't single ourselves out from the "herd."

We needn't use predatory force to get a horse to cooperate; the horse is already willing to do what we'd like, if we don't try to force it. The horse wants to find its place in the herd, but it needs to have its dignity respected.

The difference between horse whispering and "breaking" horses lies in the question, "Would you like to come and join up with me?" The horse makes a choice.

In horse whispering, attunement is an essential first step. In the round pen where he works with horses, Bob says, "Every horse is different. I listen reflectively and the horse speaks to me in its own language. I respond to the signs the horse gives me."

Just as violence begets violence, in the gentleness of joining up, being a willing partner for a horse lets the horse be a willing partner in return. Genuine change cannot be imposed. "The horse in my round pen is *disenchanted* with running and is looking for a safe way to join up with a new herd," Bob says about the horse's motivation. And Bob makes an agreeable herd companion.

The same holds for counseling clients: the interviewer creates a safe place so the client can relax into the secure mode and join up. The clinical method known as "motivational interviewing" seems parallel to this gentleness.[1] It emphasizes listening for what is important to the person and building energy for change based on those values. This means sensing the other's way of seeing and understanding, his needs, feelings, rules, and conditions, and communicating in those terms.

This suggests some basic principles for guiding anyone, whether a client in counseling or a teenager asking advice. Rather than imposing a decision, let the other choose. Keep a conversation alive and give enough time for a free choice. Be response-based, not demanding. Open the doors of opportunity rather than dictating what should happen.

When one person offers a gentle, calm, and steady center, the partner can relax and let down her guard. As with horse whispering and counseling, given room to move and some safe options, a person, like a horse, can make an intelligent, positive choice.[2]

Joining up creates an optimal environment for change and learning: you can feel seen and known, accepted as you are, and safe to risk self-disclosure or try new ways of being. The root of *trust* is the German word *trost*, which means solace or comfort—like the comforting bond of a shared secure mode. The time put into the groundwork of developing trust is always worth it because trust speeds learning and strengthens connections.

"When the horse exercises his option to come to me rather than go away, I welcome him as strongly as I can in the language of horse," says Monty Roberts, the horse whisperer credited with creating the term "joining up." "I reinforce his choice so that we work together in harmony, with the knowledge that we mean no harm to one another."

A gifted and caring bodyworker described this joined-up interconnectedness in the healing arts: "When I'm working on someone, first there's a sense of quieting and clarifying, like a gentle breeze over a still pond, so you can see what's there. You're attuning to a force that flows through all things to allow healing to happen."[3]

The Giraffe and the Jackal

Giraffes have very large hearts, they say in the nonviolent communication approach.[4] A "giraffe" person doesn't care about who is right or wrong but what matters to people—their feelings and needs. A giraffe just listens and tries to understand.

Sophie Langri teaches nonviolent communication to kids in Montreal schools. She has a natural way of using these skills. Kids, Sophie says, are natural giraffes. Kids know their needs, especially in the early years. When she asks kindergartners, "Do all humans need an iPod?" "No!" they call out in unison. "Do all humans need love?" "Yes!" "Do all humans need respect?" "Yes!"

Then, as they grow, Sophie observes this capacity seem to wane. She encourages them to connect to their feelings and express their needs, not disconnect. "You might say to an older kid, 'You don't look so happy today. What's going on?' Then kids feel *oh, we're allowed to say what we need*—they're so relieved."

One little girl in a first-grade class had a mother who was very pregnant and troubled about pregnancy complications. The mother later told Sophie, "I had a problem. Things were so difficult. I cried. I asked my friends and husband for help. No one could help me."

Then one day her daughter said to her, "Mommy, you're so sad. What's happening?"

"I am sad," said the mom.

"How do you feel?" the little girl asked.

"Scared and sad."

"I think you're feeling worried," the girl said, adding, "What are your needs?"

The mother burst into tears. The daughter said, "Don't cry yet, Mommy. There's one more step. What are you going to ask your doctor now?"

"She knows the protocol," her mother later told Sophie. "Out of all of my friends and my husband, it was my first-grade daughter who was able to help me. If my daughter learned nothing else going to school but these communication skills, I'm happy."

I've found that these communication principles fit well with how whispering applies to people—especially when viewed through the lens of a mode. Often conflicts can result from the distorted perceptions and reactivity of a negative mode. Turning to your secure mode increases your chances of finding a way through the disagreement.

When we're feeling upset, the first step is to calm our own minds. When we work beyond our minds and reactions, it can change our relationship to the issues and how we communicate. For instance, we can be firm without getting angry.

As one Tibetan teacher puts it, "No calm, no kind. No kind, no clear." A calm mind lends itself to kindness and clarity.[5]

So if we don't find our secure base ourselves, we're not going to be as effective; we're likely to just keep things going on as they are. The first step is connecting with ourselves and finding out what we need.

It can take a long time to figure out what we are truly feeling and wanting, especially for those of us who have been frozen in modes where we are out of touch with our genuine core being. Mode work can unfreeze that child within, who has given away so much personal power; it's hard to heal a hidden self. But once we contact those emotions, we need to learn the fine art of expressing them skillfully and naturally.

An empathic connection with someone who triggers your mode, or with whom you disagree, is a priority. We need to remember that we're not trying to change the other person but to create a connection where all our needs get met.[6]

Contrast the openhearted giraffe with the "jackal" way of relating, which comes from our automatic habitual reactions, like when we're in the grip of a negative mode. A jackal doesn't care how what he says makes you feel; with his predatory outlook he just wants to win the argument.

The jackal approach, where we try to get other people to do what we want—regardless of what *they* want—destroys trust; this seems predator-like through the mind-whispering lens. If we engage in a monologue, that's a sign of an I–It connection. In a dialogue, we listen to hear what the other person needs.

When we disagree, we often think the other person is wrong, and so we don't connect. The idea is to transform the jackal into the giraffe: to remember that everyone needs to feel understood and valued, and to change how you speak accordingly.

Getting both parties' needs met starts with connecting in a way where we see each other more clearly. You can't assume the other person has the same perspective you do. And try to remember that there is always some positive intent behind the other person's stance—at least from that person's perspective.

Some basic principles for handling disagreements:

- Ask yourself if you have a strong need to be right, to prove your point of view to the other person at all costs. In many situations there is no fixed right or wrong but some truth to each side, depending on your viewpoint.
- It helps to listen objectively, as though you were outside the conflict, if you can. Give empathy enough time. Be sure you've listened enough to the other person to understand that person's whole position before responding with your own.
- Connect your feelings with your needs and speak from your heart, without overplanning. We can communicate with *satyagraha* ("truth force"), Gandhi's way of assertive honesty, by staying connected to what's true for us while sending a clear message; for example, being firm without being angry.
- Be precise. What we say and do can have unintended meanings. Being sensitive to the possible ways others might take what we say can help us be more discerning in our choice of words.
- Be patient. Give the other person time to find his or her own secure mode, if possible, or at least to weaken the hold of a negative one. Remember, your goal is not just to resolve the conflict but to preserve the relationship.

Sometimes when Sandhi and I are working with Bob and I invite her to do something, she doesn't respond. I'll admit that I might assume she's not in the mood, or being a bit stubborn, or would rather be out grazing. But Bob will point out that I'm probably not making my request clear enough, so she understands. So I try again, striving to be more clear and precise, with clear direction. And when I do, she usually responds.

We often send emotional messages or telegraph intentions we are not aware of, creating an unintended disconnect that goes unnoticed. Along with precision in the words we choose, it's important to be attuned to the emotional messages we convey along with those words.

Combining care with giving voice to our deepest truths can enhance any relationship. If the outcome strengthens your connection, then both sides win.

Beneath the Differences

At the height of the Cold War, the Dalai Lama said that if the leaders of the United States and the USSR—bitter enemies at the time—"suddenly met each other in the middle of a desolate island, I am sure they would respond to each other spontaneously as fellow human beings."

Disharmony steals energy that might otherwise be put to good use. Finding harmony in friendships and communities can be a matter of trusting in the hidden potential for joining up in secure mode.

This seems a universal, primal instinct, particularly in times of crisis. Shared hardships and tragedies can strip away negative modes as we find ourselves instinctively trying to help each other, which reveals how connected we actually are. But it shouldn't take an earthquake or a reactor meltdown to bring people together this way.

Beneath all our differences we can still connect person to person. To bring peace to our relationships, our communities, and our world, we need to start with ourselves, by turning to our secure mode. If we engage in social change from a negative mode, harmony is less likely.

Being grounded in peace in our own minds, so we can bring it to others, "is the basis of social action," says the 17th Karmapa. "We first have to tame our own minds if we want to benefit others."

Groups of any kind can sometimes get stalled or caught up in long, unproductive meetings. One reason is that people express their emotions but fail to say (or perhaps even know) what they want from the group. Saying clearly what you want speeds resolution.

We may need to express our own pain with what the other person is doing that doesn't meet our own needs. This applies whether we are engaging our partner, our children, or are working for social change.

A principle from horse whispering applies: don't make the other person "wrong." As social psychologists tell us, if we see some group as "evil," we will find it hard to connect with them to begin a dialogue. We may be able to see what we don't accept as due to a mode, not the person. Still, this does not condone harmful behavior.

We're most likely to see the other person as an "enemy" or "wrong" when we are angry with that person. Angry minds lead to violent action and a violent world—and a loss of clarity and peace of mind, let alone the wisdom needed to make clear decisions. Just as the person who feels compassion is the first beneficiary, the one who feels anger is the first victim. "Resentment," says author Frank McCourt, "is like taking poison and expecting the other guy to die."

Even so, anger driven by a sense of concern can be beneficial. We can try to see anger as a signal, a warning that attention needs to be paid, and try to stay connected. And sometimes simply being heard can be a powerful message in itself.

In seeing anger clearly, it helps to separate the trigger from how we interpret what someone has done. If someone has failed to respond to a request we made weeks ago, for example, that's what she has *done*. If we say, "She was thoughtless," that's how we *interpret* what she has done. The trigger is how we perceive an act, not the act itself.

We feel as we do because of what we tell ourselves about the other person. Our mental algorithms create our lens of perception. So we can sometimes get to the root of someone's anger by asking that person to share her thoughts about what happened. Judgments cause anger, and becoming aware of those judgments may give us some clues as to how to begin to heal rifts.

Marshall Rosenberg, who developed the nonviolent communication approach, offers these key points: Live from your heart; put your judgments aside from your observations; get clear about your own needs and

feelings; make clear requests; support life-enriching connections.[7] And, I would add, all that comes more naturally from our secure mode.

Sayadaw U Pandita, my Burmese teacher in the insight meditation tradition is also the teacher of Aung San Suu Kyi. He emphasizes the importance of having what he calls "spiritual intelligence" at the individual and collective levels. This means staying true to the principles of nonharming, which protects society against the forces of greed, hatred, and delusion.

A Revolution of the Heart

While miscommunication fans the flames of discontent, and while problems may be worsened by the distorted perceptions of a mode, there are social, economic, and political realities where solutions lie at the collective level.

For a horse, the herd is its secure base. For us, the human herd operates in the same way, offering a safe place where we can feel the bond of trust that lets the secure mode thrive. Joining up together in a secure mode creates a shared feeling of comfort and reassurance, where even creative inspirations can flourish.

What needs to change in our lives and our communities to make these qualities more accessible?

"Everyone needs to be listened to, understood, and have a sense of belonging." Those words were spoken by Wilbert Rideau, who at the age of nineteen was convicted of a felony and then finally paroled after forty-two years in a Louisiana prison, when a later trial found him guilty of a lesser charge for which he had already served time.

During the years he had served, he learned to read and went on to help other prisoners find a voice by founding a prison newspaper. He now crusades for better conditions for prisoners.

Rideau spoke of his feelings of deep remorse for what he had done and the agony he went through: "I emerged from my cocoon of self-centeredness and connected to a common humanity that eventually brought me peace."

Rideau was speaking to the Dalai Lama at a conference on the power of nonviolence, held in downtown Newark, New Jersey, a city

sadly plagued by violence.[8] Each presenter had endured some form of hardship—including physical, emotional, or sexual violence, serious drug or alcohol abuse, or surviving war zones.

But each one was now a peace activist in some way, at levels ranging from the family and the community to within groups and environmental activism. The overall theme: inner peace promotes outer peace. Many told stories of redemption and recovery on their healing paths—paths that had led all of them to a dedicated practice of helping others heal their pain from similar hardships.

Sometimes research suggests that people who endure extreme hardship in their early lives find it more difficult to feel compassion for others. Perhaps this might be true for those who haven't done much inner work and healing from the hardship. But I feel quite the opposite is true when we have done so. Our own pain can give us more empathy for the pain of others and a stronger urge to help someone who is suffering.

This was clearly the case in the Newark symposium, where you could see how during their healing journeys, each person had created an inner safe haven and had found innovative ways to provide shelter to others with stormy lives. They were connecting with an inner secure foundation from which to rebuild their own lives and were dedicated to helping others rebuild their secure foundations.

They saw, too, that they had a choice. One choice was simply to fall prey to the control of their destructive habits, which may have once helped them survive extreme difficulty: hardened attitudes, self-abuse, intoxicated oblivion, helplessness, carelessness, or apathy. The other choice was to use that emotional force, impassioned by a will to be free, to restore their lives.

As a girl in Southeast Asia, Somaly Mam had been sold into prostitution but then escaped. An amazing woman, she now provides a safe haven for young girls who have fled the same plight, an imprisonment in brothels. She is rebuilding a sense of safety, re-parenting them, doing everything she can. This has been life-changing for these girls, helping them find their own secure base.

"It might take just five minutes to save them but five years to heal them," Somaly Mam has said about the girls she works with. "Love is

first. We need everything to heal them: therapy, meditation. We try to build a loving family and give love and hope by creating a safe, trusting community. Peace begins in the mind."

The same theme of rebuilding the capacity for a secure mode was affirmed by David Kerr, who founded Integrity House in Newark in 1968, now a network of treatment centers for substance abuse. From a mode perspective, substance abuse can be seen as an extreme of attachment and a desperate, misguided attempt to create a secure sense of being.

At Integrity, Kerr says, "We keep it real. If manipulation starts, we put a stop to it. Addiction does not allow peace in families. We don't tell anyone what to believe in, but we ask people to connect with a power greater than themselves"—an inner secure mode.

"Ultimately, peace must develop inside," echoes Shirin Ebadi, an Iranian lawyer who won the Nobel Peace Prize for her courageous defense of women and children, and for advocating an interpretation of Islam compatible with democracy and human rights. "Peace starts with ourselves, then it goes out to our families, and from our families to our communities, and then to our global communities."

The Dalai Lama notes "peace is more than the absence of conflict. Real peace is based on nonviolence. Deliberate restraint and willpower let us face adversity with nonviolence. Ultimately peace must live in each of our hearts. We must transform ourselves and our world."

It has struck me that all of these community activists are dedicating their lives to creating a secure base for people who did not have one. In such a shared secure mode we feel heard and seen, respected and safe, cared for and cared about, and we have a sense of trust and belonging.

An African American teenager I knew about had been walking with a friend when his friend was stabbed. The teen called 911 for an ambulance and waited helplessly as his friend bled to death, as was typical in their urban ghetto, where emergency vehicles routinely take way too long to respond. For a while, he considered trying to take some kind of revenge. But instead he realized that wasn't going to help anyone. And so he decided to become an ambulance driver, one who would be sure to get there in time.

As Martin Luther King III, son of the great Gandhian crusader for civil rights in the United States, observed, quoting his father, "At the center of nonviolence lies the principle of love."

Setting Intention

A woman came to a city hospital emergency room to be seen for sudden numbness and severe dizziness—potential stroke symptoms. After an extremely long wait, with no one paying attention to her, she asked a nurse to check if she was putting herself at risk by waiting so long.

The nurse fired back in an angry tone of voice, "You wait your turn!" and turned her back.

The woman slumped off to a corner and wept by herself.

Another nurse who saw all this seemed concerned about how the woman had been treated. She came over and put her warm hand on the woman's shoulder and said, "Why don't you come back with me? I'll make sure you get checked to see that you're not at any immediate risk."

The mean-spirited nurse had treated her like a number, not a person. "It was harder dealing with her negativity than with that health scare," the woman later said. But that kind nurse had shifted her into the secure mode when she so urgently needed it.

Negative and positive modes are everywhere in life, coloring how we affect each other in large and small ways during our routine encounters. They spread like a social virus. Goodwill and human warmth are shared when someone sincerely asks, "Can I help you?" But when there's a curt "Next!" and someone who is supposed to be helping you has a dismissive or condescending attitude, it spreads negativity.

Some organizations are taking this to heart and training their staff in what amounts to emotional habit change: how to respond with kindness even when a customer is demanding, irritating, or impatient. One method encourages thinking of something pleasant before responding rather than letting the response be a reaction triggered by the customer's demands. Like social viruses, modes can spread both ways—the toxic mode needn't be the winner.

Becoming a nourishing sender of positive modes can be enhanced through our intentions. Practice is not just sitting in meditation; it should pervade our lives, in whatever we do. Our minds, skills, intelligence, and abilities give us lots of power, but it's up to us whether we use these capacities in negative or positive ways.[9]

We can become kinder, softer, more aware, more capable, more helpful, and promote positive change in ourselves and for the benefit of others. Or we can head in a negative direction, harming others and ourselves. We have the choice every moment whether or not to make our lives more meaningful. It all depends on our intentions.

Chökyi Nyima Rinpoche advises, "Make the aspiration each day: I will make an effort to use every moment of my life in the most meaningful way, to make an improvement, to create change for the better."

Such intentional aspirations act as a guiding principle for our choices. The compassionate aspiration to help relieve the suffering of all beings helps us sort out what's worth doing and what's not.

Keep in mind these three steps: Set your mind on developing that altruistic quality. Be mindful. And when you are about to start an action, pause and notice your intention.

The Transformative
Power of the Arts

One of my favorite movies of all time is in *Latcho Drom,* a passionate journey through gypsy music and dance that takes us from their ancient origins in India's Rajasthan desert through their scattering to places like Turkey, Romania, and Spain (where their music and dance flowered into flamenco). The gypsies have long expressed their travails and joys through their performing arts.

In one scene somewhere in Eastern Europe, a young boy and his forlorn mother stand waiting for a train at a desolate station. Across the tracks a group of gypsy musicians are waiting for the same train. The boy sees them, walks across the tracks, hands a coin to one of the musicians, and asks him to play a song for his mother. They seem touched by the boy's sweet request and concern, and they hand back the coin.

Then the gypsies break into a cheerful song and start dancing. Even the little boy clumsily hops to the catchy rhythm, a huge grin on his face. And as the camera pans to the sad mother, we see her start to laugh and laugh at her son's adorable, playful antics. She seems transformed by the exuberance of a shared creative spirit.

Seen as outsiders in the far-ranging countries they inhabit, gypsies experience an us-and-them world. Yet the musicians and dancers in *Latcho Drom* seem to find a way to join up with each other and with their audience through a creative expression that connects them to a joy not defined by their struggles. They find a common pulse. Their arts let us see both the struggle and the separateness, yet they connect despite their differences.

Along the gypsy trail in Granada, Spain, a friend was walking through the town square late one night when he happened upon a gypsy musician composing, on the spot, a heartfelt song of sorrow for his niece, who earlier that day had been in an auto accident. He was waiting to hear if she would survive. He put the anguish in his heart into that song. Two guitarists sat with him, attuning to his heartfelt melody with an empathic musical resonance, sharing his pain.

A Soulful Spirit

I've been a student of *kathak,* a traditional Indian dance, which has some similarities with flamenco in foot patterns, hand movements, and rhythm. Dance historians say both have the same roots in ancient India. I dance freestyle kathak and in the French West Indies became an accidental performer where gypsy flamenco guitarist Soley was playing at a café. I found myself moved by that inspired soulful spirit to get up and start dancing.

The feeling is contagious. People of all ages—kids, adults, elders—sometimes get up spontaneously and dance along with me or play percussion along with the beat. It reminds me of the joyful exuberance in *Latcho Drom,* and how these arts have the power to inspire the gypsy in everyone—joining up through music and dance. During carnival on that island I once danced down the streets amidst a costumed crowd frolicking to the beat of marching bands. We were all connecting with our inner gypsy. It makes me wonder what it would take to inspire a widespread contagion of this joyful creative spirit.

I was waiting on hold for a business conference call once, listening to the Muzak, which had a lively beat. Without hesitation I hopped up and started to dance some salsa movements around my office. By the time the call went through and we introduced ourselves, I was in an upbeat, playful mood, which I brought right into our conversation.

Growing up in Manhattan I was surrounded by a family of professional entertainers and teachers. The city was like a creative playground, with art museums my after-school habit. I've always felt creativity of any sort—brainstorming, lyrical prose, dance, music, visual art—helps us connect with an inner freedom, one where outside-the-box solutions to life's problems can be discovered.

The expressive arts expand the limits of our sense of ourselves. While dancing, you gracefully reach and extend beyond your ordinary physical boundaries. In music you send outward sounds and rhythms far beyond your usual range of voice. Visual artists share imagined realities; writing sparks creative inspirations beyond the author's experience.

John Bowlby observed that children only play when they are in the secure mode; any kind of playfulness signals we share that mode. Evolutionary anthropologists say that there are at least two ways for people to share a surge of the pleasure-inducing brain chemical endorphin: laughing together and making music together.[1]

Art programs in schools offer kids the chance to express the creativity of the secure mode. When New York City schools cut their funding for arts programs after 9/11, my brother—a musical activist who lived in the neighborhood—started a project to bring musical instruments to local schools near his business, the Off Wall Street Jam. He would donate instruments and organize concerts to benefit the schools' arts programs. "They cut the classes just when they needed the arts the most," he said.

Can the arts be used to spread positive feeling and educate the emotions? Aesthetic mediums have an immediacy that reaches the emotions without being censored by the rational mind. The arts and a good sense of humor have been connective forces throughout social evolution, enhancers for social bonding. Time and again I've seen how the arts can connect us to a larger dimension of our beings and then share that spirit with those around us.

Music doesn't need anything elaborate to occur. One spring day in

Manhattan a street musician was playing a motley collection of objects that are usually thrown away or can be found in a recycling bin: beat-up aluminum mixing bowls, a glass milk bottle, a large industrial-kitchen-size grain bucket, and odds and ends of metal. He was whacking them energetically with a broken stick. But his music was mesmerizing. A large, appreciative crowd gathered around him on the corner of Fifth Avenue and Fifty-Ninth Street. He transformed that piece of street into an intimate jazz venue.

Outside a local health food store one day a middle-aged man was drumming a catchy beat on a large bucket. As people left the store, he connected with them, sharing his joyful spirit whether they gave anything to him or not. In a rhythmic sequence without missing a beat he'd nod and say, "Have a good day. Have a good day," and one after another, each person would walk off smiling.

The arts reveal our common human nature behind a mask of divisive ideologies. Music and dance seem perfect vehicles for this transmission. They dodge our thoughts and target the deeper parts of us that want to move to the beat, to join up. The arts are a force connecting us across divides.

This creative connective force takes us beyond our limited small selves into what in the world of gypsy flamenco is called *tener duende* (loosely meaning "having soul" or "inspired spirit"), that moment of "olé!" when we share a contagious intensification of genuine feeling and its expression. Or, as it's put across the Atlantic ocean, "You rock!"

This inspired, creative spirit can help us tap into reservoirs of inner capacities not often available during our habitual routines. Each of us can find our unique connection to this soulful quality.

Some well-known singers dropping by an encampment for a non-violent demonstration asked, "What do you need? Blankets? Food?" One demonstrator called out, "Your music to inspire us!"[2]

If we called on this creatively inspired quality more often, it could bring benefit to our lives, our communities, and our global family. It will take such an inspired spirit to tap the inner resources we'll need for the future of our planet and for global relations. We will need heartfelt compassion and countless insights to open us to new possibilities and a greater potential for meeting the challenges of our time.

Of course we need every resource—education, environmental protection, new technologies, new economic models, health programs—but all of those resources will work better if fueled by creative inspiration and motivated by goodwill and concern for the benefit of all. One force that unleashes our vision and gives us the confidence to take innovative risks is an inner safe haven. When we share that, we find strength in creative connections as a herd of sorts, an ever-expanding tribe of humanity coming together for the common good.

The Spirit of Service

Seva is the Sanskrit word for selfless service. That spirit drove the World Health Organization's efforts to eliminate smallpox; the eradication of this fatal disease was a human triumph. Several founders of the Seva Foundation took part in that global vaccination drive. Others were volunteer cooks serving and caring for the masses at the Woodstock festival in 1969.

My old friend Ram Dass is a Seva founder; his Indian guru, Neem Karoli Baba, taught him about selfless service as a spiritual path. They intend to carry on that spirit of service in this foundation, tackling needless blindness and addressing other basic human needs in the world's poverty zones.

I was exposed to one face of this spirit of selfless service during my many trips to India, where I saw how humble families shared with a visitor the best of what they had, even when they had very little. We may not be able to help everyone, but each of us can help someone.

As I was growing up I remember my mother's warmhearted concern for people going through challenging life situations. She would look after them and do what she could to help them. In my own small way I try to give back through my work and service projects. Compassion is a reflex of the heart.

For the thirtieth anniversary of the Seva Foundation, an A-list of musical activists on a larger mission donated their time for a benefit concert in Oakland, California.[3] I offered to make flower arrangements for the dingy dressing rooms and backstage greenroom as well as two giant floral displays for the front of the stage—about thirty arrangements in total.

I have studied Japanese flower arranging, but I was clueless about how to operate at this industrial level. The enormity of this project left me trusting that some unknown forces would help me.

On an early-morning run to the San Francisco wholesale flower market, as the florist was tallying up the bill for this huge stack of flowers, he asked me, "What is this for?" I told him about Seva's service projects, which moved him to say, "Why didn't you tell me?" And he went back through the long tally sheet, discounting everything by 10 percent.

On my way back to my Berkeley hotel, I stopped on a whim at a small neighborhood shop to pick up two large vases I still needed. The florist there asked to see the flowers that I needed the containers for, so I brought him out to my rented Jeep, where flowers were spilling out the windows.

He looked at me in disbelief and asked how I planned to store and arrange all these flowers by myself. I told him I would keep the flowers in buckets of water in my hotel room and arrange them the next morning. I told him about Seva and its projects for the health of the poor and how everyone was working hard to benefit this meaningful cause. He seemed moved by the foundation's humane mission and said he wanted to help too.

He had two guys who worked for him transfer the flowers from my Jeep into his flower cooler for the night. He delivered the flowers the next morning, and then he and his staff put in lots of effort to help me with things like cutting the stems and putting all the blooms in water. And the next day they collected everything after the concert was over. When magically, just an hour or so before the concert's curtain time, I was given two tickets to the performance, which had been sold out for months, I passed the tickets on to the florist and his girlfriend. That spirit of service is contagious—meaningful fun.

How do we enhance this spirit of service? "Learning is the first step in making positive changes within yourself," says the Dalai

Lama. "Learning and education help develop conviction about the need to change and increase your commitment. Conviction then develops into determination. Next, strong determination leads to action: a sustained effort to implement the changes. The final factor of effort is critical."[4]

At a Mind and Life meeting on the ecological crisis there was puzzlement over why, when so many people say they care, so few act. Thupten Jinpa shared Buddhist principles for moving from altruistic motivation to engaged action. He asked, "Why is there a gap between knowing and doing?"

Jinpa's answer: "Because that knowledge hasn't been internalized. Only when you internalize it are your values transformed. Then you engage in action from that new state of mind and it becomes a natural habit."

This sequence, Jinpa says, applies to anything from playing an instrument to social activism—and, I'd add, to working with mindful habit change, whether individual or collective. In beginning with awareness of our aspirations, discernment of the benefits plays a key role. Mindful awareness is a reminder to keep our intentions active, which can build to dedication, commitment, and confidence. This leads to joyful enthusiasm, which energizes action.

Selflessly giving wherever there is need reflects one of the more evolved qualities of the secure, integrated mode and represents a defining trait of the mode of wise compassion. I remember being on an intensive meditation retreat with Sayadaw U Pandita when he asked about my motivation to practice. I said, "I feel that I am not only here for myself but for others as well."

He smiled and said, "I feel that way too."

This selfless compassion and natural sense of interconnectedness is a powerful force for good. My teacher, Nyoshul Khen Rinpoche, once gave a talk on *bodhicitta*, the mind of compassionate awakening, during a retreat in Santa Fe, New Mexico. We were doing lots of compassion practice in our meditation sessions. But that's not enough, he said. "We need to act on our compassionate wishes and do something that actually helps others. It all boils down to intention."

Connecting Through the Arts

For a physical embodiment of connecting through the arts, take the energetic dance of the sixty-something Indian master Chitresh Das paired with the young African American tap star Jason Samuels Smith in *India Jazz Suites*.

Chitresh Das explains how *bols*—rhythmic counts of the beat—are the language of the dance.[5] He sings the count aloud as he dances an elaborate pattern to the beat of the five pounds of bells wrapped around each of his ankles. As Chitresh dances, Jason recites the traditional rhythmic count *ta ki ta tha ka dimi*.

A jazz trio takes up the rhythm, improvising within the sixteen-beat cycle, while a trio of Indian musicians' instruments elaborate within the same rhythmic cycle. Jason jumps in with an intricate tap sequence set to the same complex beat that they are all precisely following with each of their instruments: feet, bells, tap shoes, trap drums, tabla, keyboard, sarod.

The elaborate melodies, the beat, the dance movements all build in intensity, until they culminate in a stirring finale and they all land on a mighty "*SUM!*" the first beat of the rhythmic cycle.

Then they start an exhilarating set of call-and-response—kathak and tap in conversation. Chitresh brilliantly improvises elaborate dance rhythms and breathtaking steps while Jason watches attentively, his whole body seeming to absorb the moves.

Without missing a beat, the moment Chitresh's steps end Jason explodes into a high-energy tap response exactly matching the same rhythm. And as he finishes, he folds his palms together in a respectful Indian-style salutation. Then Jason leads and Chitresh responds. Then Chitresh starts again. And so it goes for a dazzling round of improvised synchronicity.

It's all superbly coordinated, whizzing by so quickly it can only be done with exquisitely attuned mirror neurons, that radar that instantly activates in one's brain what it witnesses in someone else. These neurons mirror what's going on in the person we're with—especially their movements but even their intentions.

Our mirror neurons resonate, prepping our motor systems to respond.

For Chitresh and Jason this resonance has evolved to a fine art. And sitting in the audience, I could feel my body wanting to join in the action—a sure sign that *my* mirror neurons were stirring. The thrill felt by everyone watching this dynamic duo was mirroring the delight Chitresh and Jason radiated in their dance.

When we feel an engaged rapport with someone else—that sense of attunement at the core of joining up—our mirror neurons are orchestrating a dance as elegant as *India Jazz Suites*. The crucial first ingredient of rapport's three ingredients occurs as each person pays complete attention to the other. This tunes in our mirror neurons (and other systems of our brain's social circuitry).

The second ingredient involves another type of brain cell called an oscillator. Oscillators time our responses to someone (or something) and give us our sense of rhythm. With two people in rapt conversation, their bodies move in what looks like a choreographed dance, each responding to the other instantaneously. I'd bet the oscillators in Jason's and Chitresh's brains were working at a furious rate.

The third ingredient of rapport emerges from this physical and emotional attunement: joy. It feels great to be joined up. And joy is contagious. By evening's end the *India Jazz Suites* audience was on its feet too, roaring its approval.

Over the course of the evening these two artists from very different cultural traditions connected through the universal language of rhythm, the interplay of their mirror neurons forming a bridge across a divide. This joining up through the arts reminds us how interconnected we already are.

The arts have a way of freeing people from cultural trappings and helping us to appreciate the diversity of cultural differences rather than fight over them. If people are creating inspiring music and playing together, they might just forget about why they don't get along.

"The arts have the power to communicate messages of freedom and human rights," says Aung San Suu Kyi, who once said she was a fan of Bob Marley, the Jamaican reggae singer.

Marley's mother was a black Afro-Jamaican and his father was a white businessman from England. Marley's father rejected him, and during his childhood he also didn't feel accepted by the black community because of his mixed heritage.

He found a mission in his music: to bridge such divides. At one point a bitterly fought Jamaican election was leading to gunfights between the two opposing camps in what amounted to a civil war. Marley was invited back to Jamaica from England, where he had been living at the time, to headline a peace concert in Kingston, the capital.

At the One Love Peace Concert—attended by an unprecedented crowd of 100,000—there was an electrifying moment when Marley spontaneously called to the stage the two candidates of the warring parties, Michael Manley and Edward Seaga. To a raucous tune with a reggae beat, Marley had the two men shake hands. Then he stood between them and held their arms together in the air. All the while Marley kept singing about making peace, and the crowd roared with approval.

"Me only have one ambition, y'know," he later said, in Jamaican patois. "I only have one thing I really like to see happen. I like to see mankind live together—black, white, Chinese, everyone. That's all."

From time to time some kindred spirits gather to focus on social or environmental issues, to simply connect people, share resources, or to brainstorm in a gathering of the tribe. Their aim is to find ways to make a meaningful contribution to causes they care about. At a roundtable with one such group of environmental activists, I asked, "What do you think really motivates people to want to make a difference in the world?"

An answer came from Bill McDonough, author of the ecology manifesto *Cradle to Cradle:* "People want to do something together that is meaningful."

Victor Frankl, who survived a Nazi concentration camp and lived to found a school of existential psychotherapy, described the human capacity to find meaning even in the most dire conditions. He wrote about suffering, that "its unique opportunity" lies in the ways in which we bear this burden. "Life never ceases to have meaning under any circumstances.[6]

"To achieve personal meaning," Frankl says, "one must transcend by doing something that directs it to something, or someone, other than oneself, by giving yourself to a cause or to serve another person."

When people fight for petty ends or are blinded by ideologies, the Dalai Lama says, they have lost sight of the basic humanity that binds us all together as a single human family. But that requires a new approach to global problems. Because today we are ever more interconnected and interdependent, "we need a sense of universal brotherhood and sisterhood, an understanding that we really are part of one big *human* family."

If our sense of self should somehow expand to include all others, we'd have a world that works for everyone. But as long as people remain indifferent to the feelings and happiness of others, achieving a spirit of genuine cooperation remains difficult. Conflicts arise from a failure to understand another person's shared humanity.

As the world population grows and resources become depleted, we face tragedy. A root cause can be seen in people focusing on their selfish, short-term interests, and losing sight of the rest of the human family, of the earth, and of life as a whole.[7]

Buddhist psychology sees attachment and aggression—and the underlying ignorance—at the root of these tendencies. The antidotes are love and compassion, "the moral fabric of world peace."[8] With these can come a renewal of essential qualities, such as morality, compassion, decency, and the wisdom that transforms ignorance.

We can hope that with the right education future generations will grow up with these values. In the meantime, we can set an example in our own lives and in our own practices. Two important methods for this, the Dalai Lama says, are self-examination and self-correctives.

There is no single path to this; different people need different approaches. My hope is that mind whispering gives us one way to bring those two methods into our lives.

CHAPTER

20

Collective Whispering

The scene: a drive-through line. The cars are bumper-to-bumper. An SUV pulls up to the billboard menu, and the driver rolls down his window, but he can't quite get his mouth close enough to the microphone to order. Frustrated, he leans on his horn to get the car in front of him to move forward. But the car in front is bumper-to-bumper with the one in front of him and can't move.

Now angry, the SUV driver leans out his window and yells an insult. At that, the driver in front of the SUV—who happens to be a martial arts instructor—gets angry too. His temper flares, his mind races with fantasies of revenge on the world's rude people, testosterone and adrenaline flood his body, and the world shrinks to a small orbit of rage. He has the impulse to leap from his car, reach into the SUV, and send that rude driver to the dentist.

Then this martial arts teacher glances in his rearview mirror and sees the SUV driver's face contorted with rage. He looks at his own face: the same mask of anger. He has second thoughts and his mind goes to his t'ai chi training on maintaining equilibrium in all situations.

At that moment the car in front of him drives off, opening the way for the instructor to get to the pick-up window. As the cashier hands him his coffee, he says, "I'd like to pay for the guy behind me, the one in the SUV."

The guy in the SUV has ordered breakfast for five. The instructor pays for them all and feels pretty good as he drives away.

Then the SUV driver passes on the favor. He pays for the order of the car behind him. And for hours afterward, through a long chain of drive-up orders, driver after driver passes the favor back to the car behind. Not one person takes selfish advantage, accepting the gift but not paying for the next person. That act of kindness ripples through the day.

Consider what triggered that continuous wave of good heartedness: an intentional mode shift. In an act of inner martial art, that t'ai chi instructor absorbed the anger and intentionally transmuted it into kindness. The instructor, Arthur Rosenfeld, later called that alchemy an "act of consciousness." Maintaining balance in life is all-important—the point of meditation, breath work, and t'ai chi training. These mental disciplines can be immediate ways to join up and connect with the secure mode.

The moment we start to lose our tempers, we are faced with three doors. Number One: meet force with force and start a fight—Rosenfeld's immediate impulse. Number Two: give in to the other guy's anger and ask for forgiveness. Rosenfeld chose Number Three: a middle way, which differs in each situation, but restores everyone to a better state. As he says, "The trick is to figure out what that better way is."

In a cool, calm, and collected mode we are more ready to respond to life's honking horns; our better modes are a gift to everyone around us. Because modes, like moods, are contagious, our inner state can sway that of the people we are with.

The sum total of this interpersonal exchange can determine the overall mode of a group. As with the t'ai chi instructor, someone who can keep inner equilibrium even while others are losing it can make all the difference in restoring an off-kilter group to a more balanced mode of being.

The perceptual biases of the toxic modes at an individual level can be shared by a group. We can see these invisible forces at work not just within ourselves and our relationships, but also in our families and communities, our relationship to the earth, and globally, in how nations interact. The principles that heal at the relationship level are also helpful, along with other dedicated efforts, at the collective level.

A Universal Language

When I was in Kyoto taking some lessons at a Japanese tea ceremony school, one Saturday morning I went to explore a local outdoor market where artisans display their wares. An elderly Japanese man had put out a blanket on which there were some tea bowls.

A beautiful antique tea bowl caught my eye. I held the bowl gently, appreciating its character: *wabi*, as the Japanese say—seasoned with age.

The bowl used in a tea ceremony is supposed to have a story behind it. I wanted to ask this fellow about his bowl, but he clearly spoke no English, and except for a few tea phrases I knew no Japanese. I tried to motion to him what I wanted, drawing a bit on my training in kathak, the Indian storytelling dance form, making imaginative movements that I hoped would communicate my curiosity about the bowl's background. He didn't have a clue what I was trying to ask, but he seemed amused by my antics.

Then I realized that this was the story. I would tell about this moment when I served someone Japanese *macha* tea in that bowl. Just as there's a story behind every tea bowl, if you listen closely everyone has a unique story to tell.

New York City has always embraced a wide variety of cultures, and growing up in Manhattan I got used to meeting people from all over the world. I had best friends who were Greek, Korean, Chinese, Spanish, Norwegian, Jamaican, and French. I've always loved to travel to different countries when I could and to learn traditional arts from many cultures, such as Japanese tea ceremony, Indian classical dance, and international folk dances, as well as the meditative practices and philosophies I've studied from Burma, Thailand, Tibet, and India.

In my sunroom at home I have a basket of small dolls from around the world. I playfully tell people that they all get along even though they live in the same basket. Visitors of all ages have played with those dolls, putting costumes from one part of the world on a doll from another: the Indian Barbie lends her sari to the Danish girl; the Chinese dancer tries on the Tibetan *chuba*; the Ecuadorean family takes care of an Eskimo baby. It's all in the spirit of play—but, if you think about it, there's a larger meaning. Kindness is a universal language. Leaving behind our habitual cultural assumptions is a connective force that can help melt barriers.

When someone asks me where I'm from, I often feel like I have a little bit of all these cultures infused in my being. That's partly why I feel so perplexed when nationalities are filled with so much hatred toward one another that their conflict leads them to go to war.

Perhaps it's naïve to assume that there should be a way to overcome such differences, but we can at least try to heal divides. I feel that so much conflict is due to a lack of empathy and communication. Plus, we all perceive and interpret things so differently. Perhaps we can consider finding a skillful compassion, a wise approach to mutual understanding even in the face of disagreements.

There was a poignant moment at the airport in London. At a currency exchange booth there was a man who had just arrived from an Eastern European country, one that had had Nazi concentration camps during World War II. He told the exchange clerk, "Please visit my country. It's so different than it was at that time. The younger generation doesn't have any of those feelings that the people did who lived during that time."

What is it that solidifies the past and carries resentments from long ago into the present? It's helpful to remember that things change. People who live with the burden of previous negative actions that their countries had engaged in can be quite different. That's worth reflecting on in any relationship.

From Them to Us

An ominous message reached Dr. A. T. Ariyaratne, founder of the Sarvodaya Movement, a grassroots community-development organization in Sri Lanka. His movement was becoming too popular. It was now threatening powerful parties in the small nation's elite.

Word came to Dr. Ari (as he's popularly known) that there was a plot to assassinate him. An underworld boss named Choppe Aiyah had been paid to kill him at a lecture Ari was scheduled to give at a Buddhist center.[1] Tipped off, Ari went to the home of Choppe—nicknamed "the king of killers"—and presented himself to the crime boss.

"Choppe Aiyah," Ari calmly announced as he gazed into the eyes of the surprised thug, "I am Ariyaratne, whom you are planning to kill. Please do not desecrate that sacred Buddhist seat of learning with the blood of a beggar like me. Kill me here instantly."

Shocked, Choppe replied, "I cannot kill you." From that time on, I've heard, Choppe supported Sarvodaya and became one of Ari's admirers, calling him a respectful "Our Sir."

That courageous tactic of direct, nonviolent confrontation epitomizes the strategy Dr. Ari has taken from his model, Mahatma Gandhi. The very name of Dr. Ari's movement, Sarvodaya, says it well: in Sanskrit, the root *sarvo* means "all" or "embracing everything," and *udaya* means "awakening." The movement awakens its members to that open embrace in many ways.

Even during the ferocious Sri Lankan civil war, which pitted Buddhist Sinhalese against Hindu Tamils, people lived in peace with one another in Sarvodaya villages—not just Hindus and Buddhists, but also Christians and Muslims, coexisting in friendship and as neighbors working together for common goals.

"We have to give the power back to the people," Dr. Ari says, in the spirit of movements from the Arab Spring to the Occupy movement. Sarvodaya brings everyone together to provide health care, to put in water pumps, and to build roads and housing for villages that otherwise would lack these necessities. "Power should be placed at the village level," he says. "People should be able to govern their own affairs."

From a mode perspective, this work means establishing a shared secure base, both individually and collectively: a safe space where people who otherwise might be enemies can join up as "we" rather than as "us" and "them." "When selfishness dissolves," he says, "'us' and 'them' evaporate.

"I see Sarvodaya as what I call the chemistry of connection—modes that align and link us." He emphasized the link between individual and collective transformation. "The root causes of suffering are greed, ill will, and ignorance. Organized greed widens the income gap between rich and poor, multiplies pollution, and a host of other troubles. Organized anger becomes hatred and violence, ending in war.

"But we cannot address only one problem independently. Everything is interrelated. We need dialogue rather than force, and internal disarmament to create outer disarmament. Peace is more than not fighting."

Listening to Dr. Ari, I thought of how often the mind solidifies around "us" versus "them." Toxic modes can be shared.

The Sarvodaya Movement blends Gandhi's social principles with Buddhist philosophy, focusing on rural development and self-sufficiency. But the crucial factor is cultivation of what Dr. Ari calls "self-reliant minds."

Meditation is one dimension of the movement. I heard Dr. Ari tell a group of American social activists, "Gandhi was trying to transform himself, transform his mind. If one individual is awakened, then the family, the village, the nation, the whole world can awaken.

"Heal society through nonviolent direct action," says Ari. "Transform politics and economics. Heal the environment. For this, we first need to heal the mind, to transform our thinking. Each moment is an opportunity to be mindful of thought, of speech, of action—with less greed, less hatred, less delusion."

In the series of conversations I had with Dr. Ari I was struck by the magnitude of the transformational power of his work. At the heart of his method is freeing the mind from the modes of greed, hatred, and delusion, the root causes of suffering (as we saw in chapter 3), while encouraging the secure mode through meditation, along with encouraging the causes of happiness, wisdom, and compassion—and of course social action. As he says, "This is how to heal our world."

Collective Mode Work

There are hope-inspiring activists around the world who are, essentially, creating a collective secure base. These shared safe zones of the mind allow the crucial shift from the adversarial mode of "us and them" to the collaborative mode of "we."

That dynamic underlies the work of social psychologist Ervin Staub. As a child in Budapest, Staub was saved from the Nazis by the kind intervention of the Swedish ambassador to Hungary, Raoul Wallenberg. Wallenberg granted Swedish passports to tens of thousands of Hungarian Jews and sheltered them in buildings declared Swedish territory. This Swedish umbrella prevented the Nazis from arresting them—and so rescued them from the death camps.

Staub has dedicated his career as a social psychologist to understanding the roots of evil and how to overcome it.[2] The seeds of an us-and-them mode, Staub finds, begin with a group's pride feeling wounded; behind the anger is an underlying feeling of vulnerability.

That fits with Aaron Beck's analysis: a sense of being wronged triggers hurt and shame, which quickly turns to anger and hostility, and then to retaliation at the "enemy." The us-and-them mode, he says, reflects "primal thinking," cognitive residue from our ancient past when predators, like marauding humans, were a threat to our lives. While it once may have paid to have a friend-or-foe, predator-or-prey instant reflex, today that either–or thinking gets us in trouble. This cognitive bias can lead a group to perceive misunderstandings or mild challenges as serious offenses, which then leads to needless friction.

Just as in dealing with the predator-like mode within individuals, at the collective level there's also an opportunity to intervene by altering the way people perceive themselves, their group, or their core beliefs. We need to become more aware of the rigid thinking that gains control of our minds when we feel threatened and resist the tendency to judge people in absolute categories, particularly as "the enemy."

Predator-like entitlement means people lack empathy for those they are trying to control; personalizing the other group by recognizing their common humanity, even having compassionate concern, might pierce the idea of "them" and reach the awakened part of "us."

Such an approach is being used today in Rwanda, a decade after the ferocious tribal bloodshed between the Hutus and the Tutsis, which took a million lives. Ervin Staub is working together with people from both tribes on a hugely popular radio soap opera, which spreads the message of intertribal cooperation and urges people to speak up to oppose hate speech.

When I met with Staub, he explained that he was using the soap opera to create what he calls "active bystanders": witnesses to someone being harmed who, instead of being passive (and therefore seeming to approve), do something to oppose the harm—like Wallenberg. Whether opposing a schoolyard bully or an act of bias, active bystanders have the potential power to put a stop to it.

It might be asking other bystanders, "Isn't there something wrong going on here? Shouldn't we do something?" and then finding an action appropriate to the situation. Otherwise the collective norm of just letting the harm be done strengthens.

An active bystander influences other bystanders to act too, even if just to voice their concerns and explore alternatives, which can change the social equation. What had appeared outwardly to be okay with everyone (though privately it was not) no longer is condoned. But if there is the potential for violence, bystanders need to act in a way that does not put them at risk. Which led me to ask Staub, "How can active bystanders protect themselves?"

"Good question," he said. "That's why I encourage early prevention." The longer the harm has been going on and the more intense it becomes, the harder it is to reverse direction, he's found. So his research has focused on detecting early signs, like unjustly blaming a group for problems, and "finding effective ways to intervene at the early stages."

That, I think, seems exactly parallel to mode work: being aware of the mode in its early stages and finding ways to intervene. That's one way the mode perspective might help in the social dimension: recognizing these patterns in people's minds before they lead to actions.

Seeing the parallel to predator-like, narcissistic attitudes in modes, I said to Staub, "Sounds just like when people who are entitled have no empathy and blame others for their problems—they rarely see their own faults. They don't follow the rules or take responsibility for their actions— and there's often an underlying sense of worthlessness behind that."

"It's the same in groups," he said.

Dr. Paula Green founded a graduate program in conflict transformation, based on her years of work in peace-building. Peace-building focuses on compassion and helps each side find common ground so that they can see each other as human beings just like themselves, building a secure base they can share.

Her group, the Karuna Center for Peacebuilding, held conflict-resolution sessions in Rwanda as well as with different factions in the long-simmering civil war in the Balkans. In the dialogue circles with Bosnian and Serbian women, at first no one would even look in each other's eyes.

U Vivekananda, a German-born Theravadan monk living in Nepal (in Lumbini, Buddha's birthplace), has combined Buddhist principles with conflict resolution to settle disputes that were dividing that community. He attended a workshop that Paula Green was giving in Nepal. Afterward he reported applying those methods successfully with owners of brick factories whose emissions were eroding a World Heritage Site. He found dialogue through the differences and a remedy that worked for all parties.

U Vivekananda was telling me how Buddhist principles for transforming the mind have been very helpful in this environmental activism. Effort, patience, insight, concentration, and equanimity, to name a few, are needed to accomplish change at the societal level. Then we explored some outside-the-box insights, such as instead of just relying on official agencies to make the needed change, to look into buying from factories that make green bricks by drying them in the sun instead of using furnaces that pollute.

"If everyone did something to help the environment," as César Chávez once said, "we could really turn things around."

In psychotherapy, identifying patterns releases some of the conflict: we can recognize a pattern—rather than a person—as the cause. Realizing that there are two of us creating a pattern helps us see each other's humanity.

Inner work parallels intergroup work; joining up at the collective level might add a helpful dimension to peacemaking. As I've spoken with such peacemakers, I've seen how the mode perspective can provide a clarifying conceptual framework that can help inform their work.

From our secure mode we increasingly have the confidence and inner connection that lets us be more attentive to the needs of others. We have an inner platform for working toward changes that benefit everyone. A collective secure base can be a helpful support from which to act together.

There's a step beyond: we can target the root causes of a conflict in order to be able to find apt solutions. This means recognizing that adversaries are so often caught in the same history, within a similar web of bitter antagonism, expressing similar fears and acting on the same "ill will, acquisitiveness, and essential blindness," as one Buddhist activist put it.[3]

If we aren't aware of the underlying modes, we may resolve the issue at hand, but these symptoms are likely to come back over and over again.

Melting Us and Them

Randy's family hails from Eastern Europe, though he lives in a Hispanic neighborhood in Brooklyn these days. Blond and blue-eyed, he spent the first ten years of his life on the streets of Buenos Aires then moved to an inner-city neighborhood in Baltimore. So he speaks Spanish like a native and English like one too. Randy's unusual mix of heritages—European, American inner-city, and Hispanic—lets him belie easy stereotypes.

He likes to join pick-up soccer games in Brooklyn's city parks. One day he was playing with some Hispanic guys he'd never met before. As was natural for him, whenever he spoke to one of them during the game, he'd casually speak in Spanish. But they always replied to him in English. Somehow it didn't compute that this tall, blue-eyed blond was Hispanic.

At one point after Randy had once again said something in Spanish, one of the guys stopped suddenly and said in a tone of disbelief, "You speak Spanish?"

Randy replied, in Spanish, "I've been speaking to you in Spanish for the last half hour."

Then the other guy said to him in Spanish, "I thought you were a gringo," and melted into a wide smile.

That melt signified a shift from a subtle us-and-them attitude to a collective "we." As Randy put it, "The reason I like to do that is to break through their stereotypes and assumptions. They realize, *He looks like a gringo but he feels like one of us.*"

"When we extend our own ego to the whole group," Aaron Beck observes, "and we start to think our group is right while the other group is wrong—and we're better than them—it leads to seeing us as all good or right and them as all bad or wrong. Pretty soon we start to dehumanize them—they're not real people anymore."

As we lose empathy, the "others" aren't seen as having human feelings or human rights. In this toxic collective mode, people in a group stop taking in information about the others in an open way and instead filter perceptions through the distortions of their shared skewed lens.

The need to break through our stereotypes seems even more imperative when it comes to the many intergroup conflicts around the world. Arabs and Israelis, Hutus and Tutsis, Irish Catholics and Irish Protestants, Serbs and Croats—when all such distinctions between people are amplified by a collective mode of fear and hatred, the results are explosive.

The gap that occurs between two opposing groups—the breakdown of trust, the hardening of attitudes, the obstacles to mutual empathy, the solidifying of "us" and "them," the bitterness and negativity toward the other group—all these point to the root causes of conflict. It's as though each group, aimed at the other, falls into a grand-scale version of the punitive extreme of the aversive mode.

Collaboration

Yet the sort of communication and mutual understanding that might heal such fierce divides is more likely when people come together with the intention of working toward a shared secure mode. The mode principles for individuals and relationships can also be applied to the collective sphere, from communities to nations, and particularly to intergroup relationships.

Phillip Shaver's research with Mario Mikulincer, an Israeli psychologist, has found that when we prime our secure base, our biases evaporate and we become more tolerant. This makes a shared secure base the optimal space for working together across an us-and-them divide.[4]

So a first step in resolving any intergroup conflict might be to recognize modes that separate us and turn toward those that connect. "We are willing to extend our hand," as Barack Obama once said, "if you are willing to unclench your fist."[5]

Aaron Wolf, a hydrologist at Oregon State University, has used his expertise in the ecology of water systems to bring together people who are otherwise antagonistic toward each other. He's worked in Southeast Asia, the Middle East, and Yugoslavia. Because these sometime adversaries share common concerns about something vital to their lives—the management of a water table or a river system—they are willing to talk, he finds.

But before he gets into the details of water management with them, he first asks everyone to express the symbolic, higher meaning of water in their culture and religion. He takes them all to a shared plane of understanding so they don't start with their disagreements but with a common bond. That discussion very likely primes the positive modes of everyone in the room. He finds that the rest of the negotiations are more likely to go well after that.

Also, he often shows two maps—one of the countries involved and their political borders, the other of the same territory defined only by water and land. The second map opens eyes—and hearts. Instead of focusing on what divides them, they realize what unites them.

"We go through life with our own needs and expectations," which form our boundaries, Wolf observes.[6] If we can draw a new map of our world we can include the needs and wants of the entire community.

This shared secure mode fosters collaboration, as opposed to the usual geopolitical conflict mode, common worldwide, in which a government dams the headwaters of a river or diverts it with canals, depriving everyone downstream of desperately needed water. In some instances this could mean choking off life-supporting water—a striking example of the ways predatory control can sometimes be so destructive.

A counselor from Kuwait, who uses some of my principles and practices in her work, came to a workshop. We talked about the wars in her part of the world. "It's governments that are in conflict," she said, "not the people."

When I have the opportunity to give workshops in different countries it's always refreshing to meet in that authentic, secure base connection with others who on the surface might seem different from myself. People all around the world share the same emotions, bear the same hardships, and essentially want similar things. I'm always pleasantly surprised at how happy people are to put down the burden of their group identity. The more people understand each other's perspective—even if they don't agree—the less they are likely to fight.

From afar it's all too easy to fall into simple stereotypes, but if our understanding stops there, we miss the more subtle truths that an attunement to local politics and social realities might bring. We miss the opportunity for open understanding, which fosters true empathy—where the other feels felt and understood—and the chance to build a shared secure base with mutual cooperation.

Us

A robin and her mate nested in the tree that shades our deck, which is one of my favorite places to spend time at home. Their nest was perched on a forked branch near a sliding door that gives me access to the deck. That tree was perfect for a cozy robin's nest; the canopy of leaves protected it from being seen from above.

Soon the nest was filled with eggs, and the robin spent hours sitting on them. When noises disturbed her peace, she would fly up and away from the nest, perhaps to distract predators. I adapted to sharing the space with this new flock, and I tried to use a more distant door or go through the patio to the deck more often so as not to frighten them.

It seemed as if the robin flock accommodated us as well; even with her newly hatched chicks she went busily about, flying off for food to bring back to her babies' perpetually uplifted beaks.

One day a friend walked out on the deck unaware of the presence of the hatchlings in their nest. The robin shrieked, a sound I had never heard from her. When I mentioned this to my friend, who was knowledgeable about animals in the natural world, she said, "Oh, it's because she doesn't know me. That explains why she shrieked."

Finally one morning I saw the most robust chick hopping around the branches of the tree, seemingly testing its wings for flight. A few days later I noticed that all the babies had vacated the nest. Then the next morning I saw the lone mother robin sitting on the railing of the deck for a long time. She would sometimes look up at her empty nest, and then gaze off in the distance. I, too, felt their absence. Even though now I could come and go freely and spend more time sitting under that tree on the deck, I missed their presence and watching their activities—and even accommodating their needs.

For that short, sweet time while the nest was full, the robins and I had formed a collective secure base, a group of beings who felt serene and safe together and who (at least speaking for myself) enjoyed each other's company. We had learned to accommodate one another, accept our mutual presence, and sense what was needed.

A secure mode–based social system would be life-enriching, much like bees and flowers in a symbiotic relationship: as bees get nectar from the flowers, they pollinate them. Meeting everyone's needs—physical, emotional, and spiritual—is what motivates people in such a system.

What does a "we" look like? It's as though we were a herd or tribe, a group that cares for its members. Our relationships are interdependent, not competitive, and not predator–prey, where privileged members impose their will.

My stepson Gov and daughter-in-law Erica, who live in a New England village, once got a chilling message on their town's emergency alert system, a reverse 911 phone call. An automated voice said, "A child is missing. This is not a test . . ." and went on to say where the child had last been seen, what he'd been wearing, and what to do.

Cameron, just seven years old, had wandered away from where he and his father had been building a playhouse in the woods behind their house. It was ten minutes before his father had noticed Cameron had disappeared: just as night was falling, he called out for his son and he searched the wooded ridges, but the boy was nowhere to be found. It was the night before Thanksgiving and temperatures were going to dip into the twenties.

The father's call to 911 had brought out the local police, the state troopers, and the volunteer fire department. Soon just about everyone in this small town was mobilized. The local grammar school became search headquarters. My daughter-in-law went by there at about midnight to drop off spaghetti and brownies for the searchers. Several volunteers kept the food and coffee flowing through the night. As more villagers got word, more food kept arriving. Even the county jail donated 200 sandwiches.

My stepson went over to join a search party. Hundreds of volunteers had gathered to scour the treacherous forest terrain, with its craggy ledges and sheer drop-offs. The search continued throughout the night, with helicopters, all-terrain vehicles, and search dogs.

Just after the break of dawn on Thanksgiving morning, two locals who knew the woods well discovered Cameron, who had weathered seventeen hours in twenty-degree weather. Cameron spent a day in the hospital, but then was sent home, safe and healthy.

One volunteer who had been out searching all night said that he was probably going to sleep through his family's Thanksgiving gathering, adding, "My Thanksgiving was finding this child. That's enough of a Thanksgiving for me."

"This was a total village effort," as the town's police chief put it. "That's the beauty of a community."

Connected at the Source

When I hear the phrase "give power to the people," I relax and agree wholeheartedly. Until I reflect on the meaning and ask which *part* of the people are we empowering? Whether on the individual or the global level, it matters which aspect of human nature our choices are motivated from: from our more evolved modes or from the negative, bewildered patterns that obscure our better sides.

Are we moving in a positive direction, giving power to awareness and kindheartedness, or in a negative, distorted, insecure, self-centered direction? It's up to us. The choice is ours in every moment.

Whether at the personal or the collective level, we face urgent questions: What is out of balance? What is needed? What are some underlying causes? Where are the resources to draw on? What would motivate us? What are the obstacles to change? And what are some of the solutions?

How can we heal underlying mode patterns that disconnect us? How do we decrease obstructive modes and enhance the constructive? Can we turn to a mindful overseer in a collective secure mode?

Our challenges go beyond local symptoms to systemic problems. We need new models of change more in line with an understanding of interdependence, which would render old linear models obsolete.

There's a quality missing in global politics: wisdom. We need more wise and compassionate leaders. One key to a more successful world is enhancing compassion at every level, from family and tribe to nations and the world. That can start with each of us cultivating our best human qualities.[1]

Bob says, "A horse puts its trust in you as its benevolent leader, looking out for its safety and wellbeing." The same is true for people: when you care for them in any capacity, they put their trust in you. That trust, in return, should be treated with care too. It takes kindhearted motivation to be there for others as a benevolent leader.

World leaders such as Aung San Suu Kyi, the Dalai Lama, Martin Luther King Jr., Mother Teresa, Desmond Tutu, and Gandhi all shared this benevolent quality. It is one reason why they are so loved. People feel that these great leaders have their best interests at heart. They feel safe. The feeling is based on secure modes, unselfish kindness, wise love, and a joined-up connection.

Nelson Mandela embodies such a leader. He spent close to thirty years living in a prison cell hardly larger than two outstretched arms and doing hard labor breaking up rocks. And he came out able to forgive the people who had sent him there.

When in power as South Africa's leader, Mandela said, "This is not time to take revenge on our opponents; it only reinforces the cycle of fear between us." Figures like Mandela have much to teach other leaders.

Bishop Tutu and the Dalai Lama are members of a select group called the Elders, which is composed of people who have won the Nobel Peace Prize and others who are intentionally trying to bring wise leadership to the world. "Our aim is to bring a basic transformation into society with compassion," says the Dalai Lama.

"No one," Nelson Mandela says, "is born hating another person because of the color of his skin or his background or his religion. People must learn to hate, and if they can learn to hate, they can learn to love, for love comes more naturally to the human heart than its opposite."

When we are young our very survival depends on the kindness of other people, and we spontaneously appreciate affection. But sometimes as we grow up and feel independent, we can fall into thinking *I'm superior, I can be a bully,* or *I can exploit others.* Such misconceptions lead to a radical shift, where we lose sight of the ways in which we are always connected to humanity. That extreme focus on the self can breed disaster.

It's the distinctions "us" and "them" that lead to the logic that allows for war, for destroying your neighbor as an enemy. But that attitude has been out of touch with a growing reality, as "others" become intricately connected with "us" around the globe. Then destruction of a neighbor becomes destruction of us.

We are each individual waves bobbing in a vast ocean of connection in an interdependent world. The webs of interdependence enfolding our lives range from the global economy to the tangle of supply chains that draw parts and materials together from hundreds of places in the phone we hold to our ears, and the labors of people in a thousand places that create the foods in our pantries.

The global systems that sustain life on the planet interconnect with local ecosystems, which define the conditions of our daily routines. At the subatomic level, physicists speak of "quantum entanglement" or "spooky action at a distance," where measuring the spin of one electron in a pair means the other takes on the complementary spin, even when they are separated by a great distance.

While we tend to blame our misfortunes and problems on single causes or rush to blame other people, a more holistic perspective lets us look from many directions, not just one. If we need to put "blame" anywhere, it's on the patterns and their underlying delusions.

Every event we experience, as the Dalai Lama observes, comes about because of "countless different causes and conditions, many of which are beyond any individual's control, and some of which may even remain hidden altogether."[2]

Relations of mutual dependence, rather than isolated entities, define our world. We can think of compassion practice within the framework of interdependence: as we become more loving toward others, they in turn are more likely to be more openhearted with us.

We are social animals. No matter how strong or cunning we are, we can't survive in isolation. Our future will always be tied to our neighbors and our communities, our region and the entire planet. We all need each other.

How Did We Get Here?

I was reading about how Marshall Rosenberg brings nonviolent communication into prisons to use with violence-prone inmates, like gang members. I was feeling more encouraged about the possibilities of doing conflict resolution, that perhaps there are constructive paths even people who have been violent might follow.

Then I was reading about violent conflicts around the world, and saw a graphic video of a police officer needlessly pepper-spraying the eyes of a young college student who had been peacefully demonstrating. Moments before I had been inspired by the hope of progress against resorting to violence, but after watching this scene my heart sank.

Right after that I went to my session with Bob and my horse, Sandhi. I shared these feelings with him, and the three of us (including Sandhi) stood together, as if in solidarity, reflecting quietly on the poignancy of the human condition.

I thought about how much I was learning from my horse about cooperation and prey perceptions and habits, and from Bob about predator–prey dynamics, and I mulled over the image of the officer using those predatory tactics on a peaceful demonstrator.

The pervasiveness of the bewildered modes that guide heartlessness weighed on my mind. After a silent pause, Bob said reassuringly, "You're doing what you can in your work, trying to make people aware of these patterns and to suggest alternatives. And I'm doing my part in my own way."

Not that we have answers, but we are learning how to ask and listen.

As I pondered the modes that seem to be at play when we lock horns in an us-and-them standoff, a montage of memories, thoughts, and images flashed by. They flipped through my mind's eye, like a sequence of film clips on the theme of modes in their lighter and darker dimensions— sometimes an intense drama, sometimes a warmhearted one:

Paula Green in the midst of mediating with two groups—bitter enemies who are coming to trust her—when someone asks the poignant question, "How did we get here?"

Aung San Suu Kyi brought back into the center of politics and honored in the country where she had been under house arrest for decades under Burma's repressive regime—and now recognized around the world.

Tsoknyi Rinpoche crossing paths with Mikhail Gorbachev, the former Soviet president. Gorbachev tells him, "You're creating inner peace while I'm working on outer peace."

As Martin Luther King Jr. put it, "The hope of a secure and livable world lies with disciplined nonconformists who are dedicated to justice, peace, and brotherhood."

When Adeu Rinpoche was telling us about his years in a Chinese concentration camp, I asked if he felt any anger at the guards. He looked surprised and said, "No, they were just doing their job!" He never lost compassion for *them*. Years later Adeu Rinpoche held a religious post in the Chinese government, and Chinese officials would come to pay their respects, get his blessing, and ask for his advice and guidance.

I was in a Chinese city near Tibet one winter to receive teachings from Adeu Rinpoche. Each morning on my way to where he was staying I would walk by the large park in the city's center where groups of Chinese would gather, even in the bitter December cold, wrapped up in their warm jackets. They would form into widely scattered groups and practice t'ai chi or do exercises or dance.

One morning I stood on the edge of a group of women dancing in the park to Chinese music broadcast over a loudspeaker. I kept a respectful distance, appreciating the creative ways they were moving together in the freezing cold. And I found myself imitating their movements as my mirror neurons transcended cultural boundaries. I tried to learn their dance.

Seeing me far off to the side and dancing on my own, several women turned around to watch me. Then they smiled and motioned for me to come over and join them. For the rest of that morning session I was right there in their midst, joined up through dance. We can connect one-to-one or people-to-people, even if governments don't agree.

In Costa Rica there's an expression: *pura vida,* pure life. The *pura vida* atmosphere reveals itself as a sense of the life force in all things, one we can open up to and align with or allow to get obscured by inner forces. While I was there giving a workshop, I had a qigong refresher with Steven Pague, and I asked him how he related to this phrase.

At that moment a line of migrating birds glided by, at an equal distance apart from one another and in a straight line, flapping their wings then coasting attuned to aerodynamics and to each other. *Pura vida* in flight.

Steven responded that when a local greets him with the Spanish phrase for "how's everything?" he senses they are not asking how are *you* but rather how's every*thing.* He responds in kind, to which they will reply, "Pura vida."

He went on to describe how we're not creating chi in this practice; we're stepping out of the way so the chi can flow, and trying to be in harmony with its natural essence. We glided through the graceful movements—gathering chi, storing, purifying, and, at the end, dissolving chi into the greater life force. *Pura vida* can be radiated toward all beings everywhere.

We all have the potential to choose our better halves, to yield to the part of our divided selves that is hardwired for altruism and to connect with our common humanity. Hopefully we will all wake up to the fact that it's not about me, or about you, or about us, or them.

The future of the species seems, quite possibly, dependent on awakening to that truth. It's always possible, I believe, for the causes of suffering to be removed. It seems imperative for each of us to contribute according to our capacities.

What human qualities have we given free rein to that foster using the most advanced technology for wars, lead us to ignore the environmental impacts of what we do, and allow the creation of so many divisions and such a pervasive sense of separateness? And what qualities are now needed to help us evolve into more civilized, caring societies, which could use our advanced technologies for the common good?

We need to challenge the complacent collective habits that drive us apart, the obstacles to a sense of interconnectedness, which are based on pretense and defense. How might we begin to shift deep patterns that foster the illusion of separateness? What if our news brought us more personal stories about people in other parts of the world and more positive stories rather than sensationalized negativity? What if we each made efforts to break out of our social milieu and spend time connecting with people different than ourselves?

Noting the failure of political thinking to create a world of peace and justice, the late Howard Zinn urged us to "look further, deeper, for guides to action . . . beyond ordinary solutions, drawing upon ancient wisdom to cope with the violence and insecurity of our time, giving us inspiration and hope."

The Way of the Bodhisattva, written by an eighth-century sage, Shantideva, has a compelling passage: "If these long-lived, ancient, aggressive patterns of mind that lead to my own suffering as well as the suffering of others, if these patterns still find their lodging safe within my heart, how can joy and peace in this world ever be found?"[3]

These long-held, learned patterns can be unlearned. We need to recognize their invisible hold and the hidden costs of perpetuating them. Habits die hard, and their secondary gains need to be acknowledged too. At what cost are we willing to set ourselves apart? For profit? Power? Greed or hatred? Control? Pride?

Investigating the costs of these habits of the mind and heart on our life force, and sustaining that inquiry until we have a clearer discernment—a collective wisdom that understands how destructive these habits can be—reveals the long-term consequences beyond any short-sighted gains. When we fail to see these habits at work in our lives and communities, we're more readily deceived by their empty promises.

These underlying patterns of bewilderment fuel misguided modes, which harm our planet and the species that live on it as well as degrade our collective goodwill. We need a shared secure base among our global community to begin to heal ourselves and our world.

Awakening Compassion

A schoolteacher in New Delhi showed her students a series of cards for discussion. On one card there were two men with guns. The teacher asked, "What do you think these men are going to do?"

One child said, "One is a Pakistani and the other is an Indian, and they are going to shoot each other. They are thinking, *If they stop it, we'll stop it.*"

Another child paused thoughtfully for a full two seconds before adding, "Well, if they become friends, the killing can stop."

A key to such transformation lies in re-perceiving the other.

At a Mind and Life meeting in India a scientist said to the Dalai Lama, "I struggle with feeling compassion toward people who are cruel."

The Dalai Lama replied, "You don't have compassion for their behavior, but you can wish that their minds be transformed."

Lovingkindness, compassion, finding joy in the joy of others, and inner equipoise are a prescription for this transformation. As one Buddhist scholar said of these qualities, "They are the great removers of tension, the great peacemakers in social conflict, the great healers of wounds suffered in the struggle of existence: levelers of social barriers, builders of harmonious communities, awakeners of slumbering magnanimity long forgotten, revivers of joy and hope."[4]

On the airplane traveling home from a Mind and Life meeting in India, on the leg from Delhi to London, I watched the monitor that displays the location of the plane as it flies over various parts of the world. I thought about our global dysfunctional family and felt a motherly compassion for all beings.

Flying over Pakistan and Afghanistan, inspired by the Dalai Lama's words, I found myself spontaneously reciting a *bodhicitta* aspiration: *May the mind of enlightenment arise where it has not arisen, and where it has arisen may it never wane, but further and further increase.*

I found myself again reflecting on the global need for more wisdom and compassion, as I flew over Russia and Eastern Europe . . . *May the mind of compassionate wisdom arise where it has not arisen. . . .*

Then stopping in London . . . *Where it has arisen may it not wane but further and further increase. . . .*

Over the Atlantic, flying through thick clouds as we circled over the East Coast of the United States . . . *May wisdom and compassion arise where they have not arisen . . . Where they have arisen may they not wane but further and further increase. . . .*

Dissolving into the Great Expanse

The immune system operates like an intelligent network, a system that learns. Its ability to protect us depends on its connectedness. "It's as if the immune system learned millions of years ago that detente and getting to know potential adversaries was wiser than first-strike responses," says the environmental leader Paul Hawken.

"The immune system depends on its diversity to maintain resiliency, with which it can maintain homeostasis, respond to surprises, learn from pathogens, and adapt to sudden changes. The ultimate purpose of a global immune system is to identify what is not life-affirming and to contain, neutralize, or eliminate it."[5]

Writing about the millions of people worldwide who are putting this spirit of service into action in small grassroots groups, Hawken says, "The gathering of social justice forces is like the human immune system. Just as antibodies rally when the body is under threat, people are joining together to defend life on earth."

A biologist says the way we think about evolution is too limited. We tend to focus on the 500 million or so years during which animals evolved—but this ignores what's happened since the beginning of evolution on our planet about 4 billion years ago. That's when independent primitive organisms first came together in a cooperative symbiosis to form the basic structure of the first cell. This is the primal prototype for all life: from "us-and-them" to a "we," in a shared secure base.

This sense of "we" embodies how everything in the universe is interconnected: in quantum physics a wave–particle duality has the potential to be everywhere all at once. This parallels joining up in the whispering tradition, where the illusory sense of self and other, us and them, evaporates into a sense of oneness.

Swimming in the gentle aquamarine waters of the British Virgin Islands, I noticed a small silver and yellow fish swimming along with me, just under my body. When I stopped, he stopped. When I turned, he turned. He seemed to mirror my every move.

This went on for about two hours. At some point my husband came over to see what was going on, and the little fish swam over to him, including him in our aquatic herd. Then the little fish would swim over to me, then back to him, then back to me.

No matter how far out to sea we would go, the fish followed right along, swimming back and forth between us, even in between our fingers and around our arms and hands. I imagined him looking happy to have new friends to play with. It was as though we were just different-looking fish, even with our snorkeling masks on.

I fell completely in love with him. I had never experienced joining up with a fish, like I had with horses. As my heart melted, my mind felt wide open. I felt a link through this little fish to the whole ocean, a sense of no separation, interconnectedness with other life-forms, with the world at large, and with a universal source.

Any sense of a separate self dissolved into this great expanse, which included everything that we are all a part of. In this sense of a universal joining up, of a vast interdependence, we are all connected at the source, despite our differences. We may seem to be separated by species or geographic distance or other differences, but we are all linked. As one physicist put it, "We are not only connected, we are inseparable."

We are all emanations of the same life force: we are born from the same elements, we breathe the same air, and we dissolve back into the same earth.

The moon's luminous glow and the intermittent stars
illuminate the expansive night sky.
Awed by nature's exquisite beauty,
awareness echoes this vast unencumbered space.
Thoughts appear as floating moonlit clouds
expressions of this open sky-like mind
reflect on the view from the stars,
and how from a star's perspective
all our personal preoccupations,
even global conflicts,
appear so small.

ACKNOWLEDGMENTS

I gratefully acknowledge the help of many beings in creating this book.

First, my teachers, who have shared with me a treasury of wisdom and clarified meanings through their skillful guidance—for this book and beyond. And for their compassionate missions, benefiting many beings throughout the world: His Holiness the Dalai Lama; my vipassana retreat master Sayadaw U Pandita; my Tibetan teachers Adeu Rinpoche, Nyoshul Khen, Tulku Urgyen and his wise family including Chökyi Nyima, Tsoknyi, and Mingyur Rinpoches. And to the wise guidance from Tsikey Chokling Rinpoche, Phakchok Rinpoche and Neten Chokling Rinpoche. Also His Holiness the 17th Karmapa and Khandro Tseringma.

My horse whispering guide Bob "RJ" Sadowski and my horse Sandhi (and her companions Lungta Bella, Yeshe, and Bodhi) who show me how much humans can learn from this revolutionary whispering tradition.

My insightful guide in schema therapy Dr. Jeffrey Young, who continues to be a mentor and friend.

Dr. Aaron Beck for his groundbreaking vision and wise leadership in cognitive therapy, and for his enthusiastic interest in integrating Eastern and Western psychologies.

For generous consultations on points of science: Al Shapere, Richie Davidson, Jeanluc Castagner, and Lara Costa.

Mark Hyman, M.D., Jon and Myla Kabat-Zinn for helping so many people, and to the Mind and Life community of scholars, scientists, and friends, for the many informative meetings and research reports that have helped to inform this book.

To those who have contributed through a story, idea, or article: Todd Lepine, M.D.; Diana Broderick; Hanuman Goleman; Steven Schwartz; Robin Merritt; Kathy Rosseau; Jody Nishman; Deb Brower; Maggie Spiegel; Mira Weil; Beth Ellen Rosenbaum; Sophie Langri; Krishna Das; Kirsten Doctor; Aaron Wolf; Diego Hangartner; Magnus Tiger-schiold; Cassandra Holden; and David Street.

Pandit Chitresh Das, my first Kathak Indian dance master, who embodies the magic of the arts, and my Kathak teacher Gretchen Hayden—something in the way she moves expresses a pure-hearted gracefulness that inspires me. Thanks to Mary Bowen and Soley. And to Rose Nisker, my dance collaborator and computer helper, who showed me how a complicated computer program can be seen as just a more intricate dance composition.

Erik Pema Kunsang for engaging dharma conversations and helping to clarify points for this book.

Those who have contributed in a myriad of ways: Diane Merritt; Rowan Foster; Matt Marian; Erica Goleman; Josh Baran; Anne Millikin; Benoit Minguy; Andy Oleuski; Tom Lesser; Rasmani Orth; Buzz and Luz; John Dunne; Jason and Injy Lew and the Vineyard community; Catherine Ingram; Stephan and Annetta Rechtstaffen for inviting us to be part of learning vacations; Trinette Wesly-Wellesley and the St. Barth community; Joel Zoss; Sununda Markus; Stephanie Grevin; Ram Dass; Jonathan and Diana Rose; and Roger Jahnke.

To those whose compassionate missions have contributed to making this world a better place and who have been an inspiration during the writing of this book: Dr. Ariyaratne of Sarvodaya, U Vivikenanda, Aung San Suu Kyi, Jackson Browne, Bonnie Raitt, Pete Seeger, Wavy Gravy and friends, Paula Green, Ervin Staub, Paul Hawken, Peter Matthieson, Larry Brilliant, Matthieu Ricard, Bernie Glassman and Eve Marko, my musical bodhisattva brother Bill Bennett, and many other kindred spirits-too many to mention here.

My editor Gideon Weil and literary agent Linda Loewenthal, for sharing this vision, their skillful mentoring, and wise counsel—as well as the genuine connection I feel with them both. Gideon has been a delight to work with and I've appreciated the exceptional qualities he brings to his role. And thanks to the wonderful team at HarperOne.

My husband, Daniel Goleman, for sharing this inner work, our love as well as our travels, learning, and service, all of which make life meaningful.

My delightful, loving family near and far, and dear friends who have been a sounding board and source of ideas and loving support: Jessica Brackman, Elizabeth Cuthrell, Diana Rogers, Joseph Goldstein, Jonathan Cott, and Richard Gere (and his dedication to help the Tibetan people and make Buddhist teachings available—some of which found their way into this book).

Jacalyn Bennett for her generous supportive friendship and offering the perfect place for me to do writing retreats on the top floor of her home in the French West Indies—where so many stories were written and creative breakthroughs came about.

And to all the wise-hearted people whom I've been fortunate to connect with through this work, who have inspired me and helped make this book come to fruition.

RESOURCES

Workshop Schedules

For workshops on mind whispering (which include essentials of emotional alchemy) and chemistry of connection, see www.tara bennettgoleman.com.

Books

Tara Bennett-Goleman, *Emotional Alchemy: How the Mind Can Heal the Heart* (New York: Random House, 2001).

HorseMindShip

Bob "RJ" Sadowski, HorseMindShip workshops:
http://peacehavenfarm.com

Schema Therapy and Cognitive Therapy Referrals and Professional Training

Jeffrey Young's schema therapy approach:
www.schematherapy.com/id201.htm

Aaron Beck's cognitive therapy approach:
www.beckinstitute.org

Insight Meditation Training Centers for Instruction and Guidance

Insight Meditation Society, Barre, Mass.:
www.dharma.org

Spirit Rock Meditation Center, Marin County, California:
www.spiritrock.org

The Insight Meditation Center of Newburyport:
www.imcnewburyport.com

Panditarama Center, Rangoon; Sayadaw U Pandita:
www.panditarama.net

For a national listing of local insight meditation centers, visit
www.inquiringmind.com

Tibetan Teachers, Teachings, and Practice Guidance

His Holiness the Dalai Lama:
www.dalailama.com

Gomde International:
www.gomde.org.uk/national-and-international-centres-p-34.html

Chökyi Nyima Rinpoche:
www.shedrub.org/calender.php?cid=2 and www.gomde.org.uk/national-and-international-centres-p-34.html

Tsoknyi Rinpoche:
www.tsoknyirinpoche.org

Mingyur Rinpoche:
http://tergar.org

Additional Information

For more on mindful habit change, blogs, musings, background information, service projects, and an expanded list of resources, see www.tarabennettgoleman.com.

NOTES

CHAPTER 1: THE LOTUS EFFECT

1. The train: David Cromer was quoted by Alex Witchel, "David Cromer Isn't Giving Up," *New York Times Magazine* (February 14, 2010): 29.

2. A mode in this sense refers to an overall organization of the major components of our minds and bodies: our cognition and how we process information; our emotions, impulses, and reactions; our motivations; and to some extent our biological patterns. Modes determine how we feel and think, what we desire and where we focus our attention, what we perceive and how we behave. Each of these systems—cognition, emotion, motivation, attention, and perception—has its distinct operation and functions, orchestrated together into the mode's particular coordination of our entire mental and physical architecture. A mode acts as a meta-system, with all these various subsystems coordinated in synchrony to achieve particular goals. See Aaron Beck, "Beyond Belief: A Theory of Modes, Personality, and Psychopathology," in *Frontiers of Cognitive Therapy*, ed. Paul M. Salkovskis (New York: Guilford Press, 1996): 1–25.

3. A mode organizes and defines the tapestry of our experience, while a mood, for example, only colors one thread of that tapestry. While a bad mood might leave us feeling a bit irritable, a bad mode grips us in far more pervasive ways: for example, we not only get angry more readily and exaggerate anything that might bother us, but even after our irritation passes, we might feel sad, ashamed, or anxious. All these fleeting moods are part of the same mode.

4. Psychologists talk about "states," like an emotional episode that lasts just seconds or minutes, or "traits," persisting stances that define us over years, decades, or a lifetime. A mode lasts much longer than an emotion, but unlike a trait, it can change. Some psychologists call modes "types," which is a way of pigeonholing people with one label or another. But to my way of thinking, a mode can be either a temporary state or a habit so persistent it seems a trait, depending on how frequently and for how long it grips us. But these are mindsets and emotional states we *visit*—they do not define

us. Modes are bigger than a single emotion or mood; they organize our emotions and moods. And they are less fixed than traits: modes can change. Grief, for instance, does not hold us in its grip forever but releases us when its work is done. Modes come in many sizes. Some can occur so regularly, over so many years, that they seem like a fixed trait of our being (though even these can transform, with the right effort). At the other end of the timescale, micro-modes take control for just a few moments or minutes.

5. Modes are bigger than any single mood; we can go through many mood changes while still in the same mode—sad, irritable, anxious, or joyous in turn. A mood determines our passing background feelings, while a mode captures our entire state of mind. It includes not just our emotions, but also related thoughts and goals, and mode-specific skews in perception and memory.

CHAPTER 2: THE WORLD OF MODES AND WHY THEY MATTER

1. The neuroscience: Charles Duhigg, *The Power of Habit* (New York: Random House, 2012).

2. Aaron Beck, "Buddhism and Cognitive Therapy," *Cognitive Therapy Today* 10, no. 1 (Spring 2005): 1–4.

CHAPTER 3: ROOT CAUSES

1. Fifth-century text: Bhadantacariya Buddhaghosa, *Visuddhimagga: The Path of Purification,* trans. Bhikku Nanamoli (Berkeley: Shambala Publications, 1976). A good contemporary source: Jack Kornfield, *The Wise Heart* (New York: Bantam Books, 2008).

2. That seminar, part of a meditation retreat, was led by Jack Kornfield, a teacher of mindfulness meditation and a psychologist, who has used this framework in his teaching. See Kornfield, *The Wise Heart*.

3. Nathan A. Fox and Bethany C. Reeb, "Effects of Early Experience on the Development of Cerebral Asymmetry and Approach-Withdrawal," in *Handbook of Approach and Avoidance Motivation,* ed. Andrew J. Elliott (New York: Psychology Press, 2008), 35–49.

   Primitive drivers of like and dislike: Alexis Faure, Jocelyn Richard, and Kent Berridge, "Desire and Dread from the Nucleus Accumbens: Cortical Glutamate and Subcortical GABA Differentially Generate Motivation and Hedonic Impact in the Rat," *PloSOne* 5, no. 6 (June 18, 2010): e11223.

4. Pavlov, in the earliest studies of conditioning, put it in terms of two reflexive responses, orienting toward something or a defensive reaction to withdraw from it. Carl Jung, in one of the first theories of personality, saw people as either extraverts, who approach the social world, or introverts, who withdraw. And the pioneering psychoanalyst Karen Horney was among the first to place this dynamic within the mind itself; she saw that some people handled their anxiety by moving toward it, others by moving

away. The mind's operations: Andrew J. Elliott, "Approach and avoidance motivation," *Handbook of Approach and Avoidance Motivation,* ed. Andrew J. Elliott (New York: Psychology Press, 2008), 3–14.

5. Michael Meaney at McGill University: Darlene Francis et al., "Nongenomic transmission across generations of maternal behavior and stress responses in the rat," *Science* 286 no. 5442 (November, 1999): 1155–1158.

CHAPTER 4: INSECURE CONNECTIONS

1. John Bowlby, *The Making and Breaking of Affectional Bonds* (London: Tavistock, 1979): 129.

2. Shaver and others who apply this model in studying relationships categorize people as falling into one or another "attachment style." I prefer to see such styles as meaning that a given person is consistently prone to having one or another mode triggered, but this does not imply a fixed way of being that we are stuck with for life. To my way of thinking, these categories describe transient states of being, not permanent fixtures of our minds (powerful though they may be). For more details on secure and insecure modes, see Mario Mikulincer and Phillip R. Shaver, "Adult Attachment and Affect Regulation," in *Handbook of Attachment,* eds. Jude Cassidy and Phillip R. Shaver (New York: Guilford Press, 2008): 503–31.

3. John Bowlby, *A Secure Base* (New York: Basic Books, 1988).

4. Secure mode indicators: the three items are adapted from examples given for items from attachment surveys in Mario Mikulincer and Phillip R. Shaver, *Attachment in Adulthood: Structure, Dynamics, and Change* (New York: Guilford Press, 2007).

5. Avoidant indicators: Mikulincer and Shaver, *Attachment in Adulthood* (2007).

6. Jeffrey Young calls this extreme mode the "detached protector." Young and First, 2003: www.schematherapy.com/id72.

7. Although suppressing distressing thoughts and feelings provides some temporary calm, in fact the anxiety can still manifest indirectly through insomnia, psychosomatic symptoms, and other health problems. The avoidant strategy can break down and fail when emotional threats are severe and persistent—as well as with overwhelming, traumatic events. Then someone who tends to the avoidant mode may suddenly gets pitched into focusing on his or her emotions in ways similar to the anxious mode.

8. Anxious indicators: Mikulincer and Shaver, *Attachment in Adulthood* (2007).

9. What Jeffrey Young calls the "vulnerable child" mode seems a variation of the anxious mode. This mode also can have an angry aspect due to frustration or rage triggered when such people feel that their needs to be protected or cared for are not being met. At the very extreme, the anxious mode includes crippling anxiety disorders, like claustrophobia.

10. Attachment style and brain mechanisms: Omri Gillath et al., "Attachment-Style Differences in the Ability to Suppress Negative Thoughts: Exploring the Neural Correlates," *NeuroImage* 28, no. 4 (2005): 835–47.

11. Matthew M. Botvinick, Jonathan D. Cohen, and Cameron S. Carter, "Conflict Monitoring and Anterior Cingulate Cortex: An Update," *Trends in Cognitive Sciences* 8, no. 12 (2004): 539–46.

12. In contrast to both the anxious and avoidant modes, women in the secure mode were readily able to quiet this circuitry for relationship angst and to turn their attention to other thoughts at will. The brain scan revealed these more worry-free women were able to activate a neural switch that calms the worry circuitry.

CHAPTER 5: AN EVOLUTIONARY ARMS RACE

1. The basic choice. For more details see *Handbook of Approach and Avoidance Motivation,* ed. Andrew J. Elliott (New York: Psychology Press, 2008). Also see "Nematode Study Identifies a Gene for Staying or Going," *New Scientist* 209, no. 2804 (March 19, 2011): 12. This gene controls secretion of adrenaline, the motor for pursuing food or fleeing a predator throughout the animal kingdom. The question "Should I stay or should I go?" reflects an ancient quandary.

2. Ian McCollum, *Ecological Intelligence* (Capetown, South Africa: Africa Geographic, 2005).

3. I–It: Martin Buber, *I and Thou,* trans. Walter Kaufmann (New York: Simon and Schuster, 1990).

CHAPTER 6: TRAPS, TRIGGERS, AND CORE BELIEFS

1. Jeffrey Young and his collaborators have captured the essence of many such destructive modes. Young's approach, schema therapy, helps find the most effective treatment for a given person. The main "supermodes" of this model each have several subtypes. There are endless varieties of refinements of these into individualized modes, each depending on the unique specifics of how a given person's life history shapes the way that person expresses them. Jeffrey Young, Janet Klosko, and Marjorie Weishaar, *Schema Therapy: A Practitioner's Guide* (New York: Guilford Press, 2003). See also Jeffrey Young and Michael First, "Schema Mode Listing" (2003): www.schema therapy.com/id72.htm.

2. Aaron Beck says that what modern psychiatry calls a "disorder" often refers to just such extreme modes. Beck points out both the modes of anxiety and of depression are extremes of normal experiences of feeling anxious or down; the "disorder" begins when we cannot extricate ourselves from these modes and they intensify to the point that they keep us from living effectively. Aaron Beck, "Beyond Belief: A theory of modes, personality, and

psychopathology," in Aaron Beck and Paul Salkovskies (eds.) *Frontiers of Cognitive Therapy* (New York: Guilford Press, 1996): 1–25.

3. David Whyte, "My Poetry," in *Everything Is Waiting for You* (Langley, Washington: Many Rivers Press, 2003): 9–10.

4. Fight-flight-or-freeze: Jeffrey Young and Janet Klosko, *Reinventing Your Life* (New York: Penguin, 1994).

5. Young calls this the overcompensating mode.

6. Jeffrey Young had identified about a dozen such schemas at the time I worked and studied with him at the Cognitive Therapy Center of New York. By now Young has named around twenty. See Young, Klosko, and Weishaar, *Schema Therapy* (2003).

7. Thanks to Mark Washington's grandmother.

8. Algorithm: Aaron Beck, "Beyond Belief: A theory of modes, personality, and psychopathology," in Aaron Beck and Paul Salkovskies (eds.) *Frontiers of Cognitive Therapy* (New York: Guilford Press, 1996): 1–25.

9. Thanks to Krishna Das.

10. I've described these schemas in more detail in my book *Emotional Alchemy* (New York: Harmony Books, 2002); see also Young and Klosko, *Reinventing Your Life* (1994). The schema list was developed by Jeffrey Young and colleagues; I've adapted it here.

CHAPTER 7: THE EVOLUTION OF EMOTION

1. For example, see F. G. Lopez, "Adult Attachment Security: The Relational Scaffolding of Positive Psychology," in *Oxford Handbook of Positive Psychology* (New York: Oxford Univ. Press, 2009).

2. Young and First, "Schema Mode Listing" (2003): www.schematherapy.com/id72.htm.

3. Tulku Thondup makes this point.

4. Playful while awakening: the 17th Karmapa.

5. Bob told me this story.

CHAPTER 8: SHIFTING THE LENS

1. Psychosclerosis: thanks to Ken Dychtwald.

2. Aaron Beck's original concept of modes focused on conditions like depression and PTSD (though such clinical modes are not included in the ones I deal with here). And as a clinical psychologist will tell you, each such mode has specific, tailored shifters—systematic ways to replace the negative routine with more positive responses to the same cues.

3. Joseph LeDoux, *The Emotional Brain* (New York: Simon and Schuster, 1996).

4. Likewise, the various sciences each have their own lens on the choice. Evolutionary psychology sees the divide as between adaptive and maladaptive

ways of being. Developmental psychologists speak of insecure and secure modes. Neuroscientists talk of functional and dysfunctional brain states, while clinicians see duality between adjustment and pathology.

CHAPTER 10: THE MINDFUL OVERSEER

1. William Faw, "Pre-Frontal Executive Committee for Perception, Working Memory, Attention, Long-Term Memory, Motor Control, and Thinking: A Review," *Consciousness and Cognition* 12, no. 1 (March 2003): 83–139.

2. Elkhonon Goldberg, *The Executive Brain* (New York: Oxford Univ. Press, 2001): 24. The executive circuitry offers a larger perspective and the ability to test reality. It can make long-term plans, change our strategies, and set new goals.

3. A wiser choice. See S. M. McClure, et al., "Separate Neural Systems Value Immediate and Delayed Monetary Rewards," *Science* 306 (2004): 5796–804.

4. Brooding versus reflection: Wendy Treynor, Richard Gonzalez, and Susan Nolen-Hoeksema, "Rumination Reconsidered: A Psychometric Analysis," *Cognitive Therapy and Research* 27, no. 3 (June 2003): 247–59.

5. Reflection versus brooding: Monika Ardelt, "How Wise People Cope with Crises and Obstacles in Life," *ReVision* 28, no. 1 (2005): 7–19.

6. Kevin Ochsner, Silvia Bunge, James Gross, and John Gabrieli, "Rethinking Feelings," *Journal of Cognitive Neuroscience* 14, no. 8 (November 2002): 1215–29.

7. My late teacher Adeu Rinpoche said this.

8. Tsoknyi Rinpoche made this point to me.

9. A Tibetan teacher: Tsoknyi Rinpoche.

10. Freud called the psyche's executive our "ego." Aaron Beck calls the mind's overseer the "conscious system." Beck notes that this system can be seen as holding our conscious sense of identity, intentional choices, will, and values, even our aesthetic sensibility.

CHAPTER 11: MODE WORK

1. Aaron Beck and the Dalai Lama dialogued at the Sixth International Congress of Cognitive Therapy, Göteborg, Sweden, June 2005.

2. Aaron Beck, "Beyond Belief," *Frontiers of Cognitive Therapy* (1996): 17.

3. Quote from the Dalai Lama.

4. For more information about schema therapy, see Young and Klosko, *Reinventing Your Life* (1994). For interested professionals, see Young, Klosko, and Weishaar, *Schema Therapy* (2003). To find a schema therapist in your area, contact the Cognitive Therapy Center of New York (www.schema therapy.com/id199.htm). Not all schema therapists make the integration with mode mindfulness or the related methods of mode alchemy, so you may need to bring these to the work of therapy on your own.

5. Inner reparenting: Young, Klosko, and Weishaar, *Schema Therapy* (2003).

6. Arguments: John Gottman, *Why Marriages Succeed or Fail* (New York: Simon and Schuster, 1995).

CHAPTER 12: PRIMING OUR SECURE BASE

1. Hermann Hesse, *Siddhartha* (New York: New Directions, 1951): 87.

2. It was clear my first boyfriend and I had grown in different directions, even though he still had a special place in my heart. Loving someone doesn't necessarily mean sharing a life with them.

3. No doubt the same effect would occur with any romantic partner or loved one. J. A. Coan, H. S. Schaefer, and R. J. Davidson, "Lending a Hand: Social Regulation of the Neural Response to Threat," *Psychological Science* 17, no. 12 (2006): 1032–39.

4. The Who-To Questionnaire: C. Hazan and D. Zeifman, "Sex and the Psychological Tether," in *Advances in Personal Relationships: Attachment Processes in Adulthood,* eds. K. Bartholomew and D. Perlman (London: Jessica Kingsley, 1994): 5:151–177.

5. Secure mode primes: See, e.g., Mario Mikulincer and Phillip R. Shaver, "Attachment Theory and Intergroup Bias: Evidence that Priming the Secure Base Schema Attenuates Negative Reactions to Out-Groups," *Journal of Personality and Social Psychology* 81, no. 1 (July 2001): 97–115.

6. Barbara Fredrickson, Michael Cohn, Kimberly Coffey, Jolynn Pek, and Sandra Finkel, "Open Hearts Build Lives: Positive Emotions, Induced Through Lovingkindness Meditation, Build Consequential Personal Resources," *Journal of Personality and Social Psychology* 95, no. 5 (November 2008): 1045–62.

7. Marc Berman, Jon Jonides, and Stephen Kaplan, "The Cognitive Benefits of Interacting with Nature," *Psychological Science* 19, no. 12 (2008): 1207–12.

8. Aaron Beck, "Buddhism and Cognitive Therapy," *Cognitive Therapy Today* 10, no. 1 (Spring 2005): 1–4.

9. Tsoknyi Rinpoche teaches this method for dealing with *lung*.

10. Lovingkindness in Richie's lab: "Latest Findings in Contemplative Neuroscience." Presented by Richard Davidson at Mind and Life XXI, Madison, Wisconsin, May 11, 2010.

11. Connectedness: James J. Gross, "Emotion Regulation," in *Handbook of Emotions*, 3rd edition, eds. Michael Lewis, Jeannette Haviland-Jones, and Lisa Feldman Barrett (New York: Guilford Press, 2008): 497–512 .

12. S. McGowan, "Mental Representations in Stressful Situations: The Calming and Distressing Effects of Significant Others," *Journal of Experimental Social Psychology* 38, no. 2 (March 2002): 152–61. Those prone to the anxious mode are less able to prime an inner secure base just by bringing to mind soothing figures in their lives than are those who more regularly find themselves in the secure stance.

13. I adapted lovingkindness to work with schemas when I wrote *Emotional Alchemy* (New York: Random House, 2003), and here for mode work.

CHAPTER 13: TRAINING THE MIND

1. Depression and panic disorder are described as modes by Aaron Beck, "Beyond Belief," *Frontiers of Cognitive Therapy* (1996).
2. U Tejaniya tells the tale of his depression and how he recovered in an interview with James Shaheen, "The Wise Investigator," *Tricycle* (Winter 2007): 44–47.
3. Adeu Rinpoche made this point.
4. Allows more room: an insight from Erik Pema Kunsang.
5. I first met U Tejaniya when he was teaching at the Insight Meditation Center, and I became interested in his unique way of enhancing awareness of the root causes of attachment, aversion, and delusion.
6. This mindfulness exercise is based on instructions from Sayadaw U Tejaniya. For more see Ashin Tejaniya, "Awareness Alone Is Not Enough," "Don't Look Down on the Defilements,"and "Dhamma Everywhere." All his teachings are published for free distribution, and are available at www .sayadawutejaniya.org.
7. Mingyur Rinpoche has accomplished a huge amount, although he's just in his thirties. He did his first three-year retreat when he was only thirteen, and then was made retreat master for the next three-year retreat. He's written two books, built a large monastery in Bodh Gaya, and designed an elaborate dharma curriculum for Westerners. As I write this, he's just left for another three-year retreat, this time as a wandering yogi. He took nothing with him save his passport and an extra robe.
8. Objectless shinay meditation: Yongey Mingyur Rinpoche, *The Joy of Living* (New York: Harmony Books, 2007): 138–141.
9. Mingyur Rinpoche, *The Joy of Living*, 141.

CHAPTER 14: WISE HEART

1. Chogyam Trungpa, *Smile at Fear* (Boston: Shambhala, 2009).
2. Mark Epstein is among the leading thinkers.
3. The six paramitas teaching came from Chökyi Nyima Rinpoche.
4. For more details of Adeu Rinpoche, his teachings, and his time in prison, see Adeu Rinpoche, *Freedom in Bondage: The Life and Teachings of Adeu Rinpoche,* ed. Marcia Binder Schmidt, trans. Erik Pema Kunsang (Kathmandu, Nepal: Rangjung Yeshe Publications, 2011).
5. Signs of progress: Padmasambhava, *Teachings from Laurel Ridge*, trans. Erik Pema Kunsang (Berkeley, CA: Rangjung Yeshe Publications, 2009).
6. Dalai Lama, 2011, *Beyond Religion: Ethics for the Whole World* (New York: Houghton Mifflin): 73.
7. H. H. Dilgo Khyentse Rinpoche, *Maha Ati* (Yeshe Melong Publications,

P.O. Box 514, Mt. Shasta, CA, 1994).

8. Thanks to Magnus in Sweden.

9. Prayer: The Four Dharmas of Gampopa, slightly paraphrased.

CHAPTER 15: THE PHYSICS OF EMOTION

1. Buddhist psychology details links in the chain leading from initial perception to action. At the seventh link, for example, the positive or negative sensations arise that are aroused by the object of awareness. At the eighth comes the desire to possess or grasp. At the ninth comes the action that follows.

2. Primitive drivers of like and dislike: Alexis Faure, Jocelyn Richard, and Kent Berridge, "Desire and Dread from the Nucleus Accumbens: Cortical Glutamate and Subcortical GABA Differentially Generate Motivation and Hedonic Impact in the Rat," *PloSOne* 5, no. 6 (June 18, 2010): e11223.

3. Dzigar Kongtrul Rinpoche, quoted in Pema Chödrön, *Practicing Peace in Times of War* (Boston: Shambhala, 2006), 56.

4. Tibetan Buddhism speaks of eight hopes and fears that solidify our sense of self—that is, hook us. The four fears are of pain, whether mental or physical; of loss; of blame; and of getting a bad reputation. The hopes are for pleasures, gain, praise, and fame.

5. That insight into impermanence is a powerful way to develop the sort of "falling away" or disenchantment with *samsara* that furthers spiritual progress. Pali has a special word for this: *samvega,* feeling deeply moved by contemplating life's dissatisfactions, and so being invigorated to pursue spiritual practice.

6. While existential philosophy uses the term "emptiness" to mean a bleak sense of meaninglessness, in the East the term has an entirely different sense. The Dalai Lama has said that "emptiness is very full," meaning that once we drop our misperception of entities as separate, we can sense the endless chain of cause-and-effect that interconnects everything.

7. Thich Nhat Hanh, *Fragrant Palm Leaves: Journals 1962–1966* (New York: Riverhead Books, 1966).

CHAPTER 16: TWO OF ME, TWO OF YOU

1. Jackson Browne's lyric inspired my idea of how we all harbor "two of us." When I mentioned this to Jackson he seemed intrigued by this adaptation but said that this song had a more personal meaning for him at the time he wrote it. Introspective songwriting like this is similar to good storytelling: anyone can relate to it and derive their own meaning.

2. Five-to-one: John Gottman, *Why Marriages Succeed or Fail* (New York: Simon and Schuster, 1995).

3. Elizabeth Gilbert was interviewed in *Body+Soul* (January/February 2010): 24.

4. Ann Gruber-Baldini, "Similarity in Married Couples: A Longitudinal Study of Mental Abilities and Rigidity–Flexibility," *Journal of Personality and Social Psychology* 69, no. 1, (1995): 191–203.

5. Or, as Mary Ainsworth put it, the nature of your attachment formed with your primary caregiver primes your expectations for later affectionate bonds.

6. Unpredictable people: Julianne Holt-Lunstad et al., "Social Relationships and Ambulatory Blood Pressure: Structural and Qualitative Predictors of Cardiovascular Function During Everyday Social Interactions," *Health Psychology*, 22, no. 4 (2003): 251–59.

7. Constant stress: Carol Ryff and Burton Singer, eds., *Emotion, Social Relationships, and Health* (New York: Oxford Univ. Press, 2001).

8. Carl Whitaker.

9. Thich Nhat Hanh, *True Love* (Boston: Shambala, 2006): 4.

10. KD is better known as Krishna Das, a renowned singer of Indian devotional hymns.

CHAPTER 17: JOINED AT THE HEART

1. Daniel Siegel, *The Mindful Brain* (New York: W.W. Norton, 2007).

2. I–You: Martin Buber, *I and Thou*, trans. Walter Kaufmann (New York: Simon and Schuster, 1990).

3. Joanna Macy calls this "linear causality." See "Allegiance to Life: An Interview with Joanna Macy," *Tricycle*, Summer 2012.

4. Perry Wood, *Secrets of a People Whisperer* (Berkeley, CA: Ulysses Press, 2005).

CHAPTER 18: A SHARED SECURE BASE

1. William Miller, *Motivational Interviewing* (New York: Guilford Press, 2002).

2. For an adaptation of horse whispering to humans, see Perry Wood, *Secrets of a People Whisperer* (2005).

3. Thanks to Jody Nishman.

4. Marshall Rosenberg, *Non-Violent Communications* (Boulder, CO: Sounds True, 2012).

5. Chökyi Nyima Rinpoche.

6. Marshall Rosenberg, *The Surprising Purpose of Anger* (Encinitas, CA: PuddleDancer Press, 2005), 21.

7. Marshall Rosenberg, *Surprising Purpose of Anger* (2005), 7.

8. Wilbert Rideau participated in the Newark Peace Education Summit: The Power of Nonviolence held in Newark, New Jersey, May 13–15, 2011.

9. Chökyi Nyima Rinpoche made this point.

CHAPTER 19: THE TRANSFORMATIVE POWER OF THE ARTS

1. Endorphins: Robin Dunbar, "Social Networks," *New Scientist* 2859, (April 3, 2012): 28-29.
2. The singers: David Crosby and Graham Nash, at the Occupy Wall Street encampment .
3. The list of musical activists included: Jackson Browne, Bonnie Raitt, Elvis Costello, Graham Nash and David Crosby, among others. Also Wavy Gravy.
4. Dalai Lama, *The Art of Happiness* (New York: Riverhead Books, 1998): 219.
5. Chitresh Das was my first Indian dance teacher; his centers are in the San Francisco Bay Area, where he teaches kathak yoga and dance, and in Bengal, India.
6. Victor Frankl, *Man's Search for Meaning* (Boston: Beacon Press, 1992).
7. Dalai Lama, *A Human Approach to World Peace* (Boston: Wisdom Publications, 1984): 8.
8. Dalai Lama, *A Human Approach to World Peace,* 11.

CHAPTER 20: COLLECTIVE WHISPERING

1. Ariyaratne's would-be assassination is recounted in Catherine Ingram, "The Gandhi of Sri Lanka: An Interview with A. T. Ariyaratne," *Inquiring Mind*, (Summer 1985): 8-10.
2. Ervin Staub, *The Roots of Evil: The Origin of Genocide and Other Group Violence* (New York: Cambridge Univ. Press, 1989).
3. Ken Jones and Kenneth Kraft, *The New Social Face of Buddhism: A Call to Action* (Boston: Wisdom Publications, 2003), 146.
4. Mario Mikulincer and Phillip R. Shaver, "Attachment Theory and Intergroup Bias: Evidence that Priming the Secure Base Schema Attenuates Negative Reactions to Out-Groups," *Journal of Personality and Social Psychology* 81, no. 1 (July 2001): 97–115.
5. Spoken in his inaugural address, January 20, 2009.
6. Aaron Wolf in: Daniel Goleman and the Center for Ecoliteracy, *Ecoliterate: How Educators are Cultivating Emotional, Social, and Ecological Intelligence* (San Francisco: Jossey-Bass, 2012).

CHAPTER 21: CONNECTED AT THE SOURCE

1. The Dalai Lama makes this very point.
2. Dalai Lama, *Beyond Religion* (New York: Houghton Mifflin, 2011): 79.
3. Shantideva, quoted in Pema Chödrön, "Practicing Peace in Times of War:" 14.
4. Nyanaponika Thera is quoted in Padmasiri de Silva, *The Psychology of Emotions in Buddhist Perspective* (Kandy, Sri Lanka: Buddhist Publication Society, 1976): 28.
5. Paul Hawken, *Blessed Unrest* (New York: Viking, 2007): 143–144. Hawken was drawing on the insights of Francisco Varela.

INDEX

abandonment, fear of, 67, 80, 151
acceptance, 22–24
adaptive unconscious, 110
addiction, 26, 134, 257
Adeu Rinpoche, 187, 293
Afghanistan, 296
Africa, 110
Aiyah, Choppe, 277
alcoholism, 134, 257
amygdala, 33, 76, 153, 199
anger, 18, 25, 26, 44, 62, 93, 97, 209, 210, 216–17, 254, 273–74, 279
    clarifying, 216–17
"angry child," 62
anxious mode, 38, 39, 42–44, 45–46, 65, 66, 67–68, 79, 80, 82, 83, 108, 134, 144–45, 146, 154, 161, 166, 170, 171, 203, 222, 223, 224, 228
applied mindfulness, 124–26
appraisal, 114, 120–21
approach, 30–31, 32, 38, 62
Arab Spring, 277
Aristotle, 78
Ariyaratne, A. T., 276–78
arts, 113–14, 158, 261–71
    connection through, 268–71
    meditative, 183–84
    transformative power of, 261–71
aspirations, intentional, 258–59
attachment, 25–28, 31, 38, 80, 83, 95, 116, 134, 140, 146, 151, 176, 189, 198, 271
    approach and, 30–31, 32
Attender, 114

attention, 114, 121, 124, 126–27, 169
    one-pointed, 169–72
    sustaining, 126–27
attunement, 119, 120, 122, 124
authenticity, 74–75
Avatar (film), 236
aversion, 25–32, 46, 62, 65, 80, 83, 120–21, 140, 146, 172, 176, 199, 201, 205
    avoidance and, 30–31, 32
avoidant mode, 30–32, 38, 39–42, 45–46, 62, 64, 67–68, 79, 80, 83, 89, 102, 108, 147, 170, 176, 223, 224, 227–28, 241
awakening, 187–91, 213–14
awareness, 100–111, 123, 125, 134–35, 141, 172–74, 175, 178, 198, 198, 267, 289
    identifying feelings, 174–77
    physics of emotion, 195–217
    training the mind, 165–78

Baba, Neem Karoli, 265
Balkans, 281
Bangkok, 3
basal ganglia, 15, 89, 103
bayu, 101, 105, 115
Beck, Aaron, 7, 20, 129–31, 279, 283
beginner's mind, 91
being present with heart, 132–34
Bennett, Jacalyn, 277
bewilderment, 25–28, 31, 80, 83, 140, 146, 176, 213, 295
    to wakefulness from, 213–14

biology, 5
    lotus effect, 5
birds, 285–86
blood pressure, 224–25
boundaries, 136, 143
Bowlby, John, 37, 38, 39, 70, 263
brain, 11–12, 30, 33, 76, 89, 91, 93, 103,
    109, 114, 153, 197, 199, 236, 246,
    263, 268, 269
    amygdala, 33, 76, 153, 199
    fight, flight, or freeze response, 33–34
    limbic, 116, 117
    modes in the, 45–46
    neuroscience of habit, 15–17
    positivity and, 76, 78
    prefrontal circuitry, 76, 103, 114–15,
        116–17, 121
    "what-if" cortex, 45
breathing, 170, 171, 274
    mindfulness of, 104–5, 170–72
British Virgin Islands, 30, 158, 297
Broderick, Diana, 104–5
Browne, Jackson, 221
Buddhism, 4, 9, 10, 18, 25, 26, 31, 38,
    71, 80, 118, 121, 122, 151, 161, 162,
    165–68, 175, 181–89, 193, 196, 197,
    200–203, 208, 209, 239, 253, 267,
    271, 277, 278, 281, 295, 296
Burma, 10, 165–66, 200, 255, 275, 293

calmness, 5, 125, 136, 152–53, 162,
    168–69, 177, 251
calming the mind, 168–69
Catholicism, 76
cause-and-effect sequences, 196–98, 201–3
central nervous system, 34
chain of dependent arising, 195
change, 87–97, 133, 151
    habit, 88–92, 108–10, 115, 142–44,
        213–14
    shifting the lens, 87–97
Chartres Cathedral, 179, 180
Chha, Achhan, 160
childhood, 62, 63, 64, 69, 70, 72, 97,
    132, 149, 224, 231, 233–34, 238,
    245, 250
    secure, 159–60

China, 187, 293
chocolate mindfulness, 198–99
Chödrön, Pema, 203
choice, 95–97
    wise, 102–4
Christianity, 182, 277
clarity, 5, 12, 18, 20, 48, 80, 152–53,
    162, 171, 173, 181, 213, 216–17
    anger and, 216–17
clinging, 26, 31, 43, 46, 65, 116, 145,
    196, 204, 207, 228, 239
co-evolution, 52
cognitive empathy, 246, 247
cognitive science, 197, 208
cognitive therapy, 7, 120–21, 129–30
Cohen, Leonard, 74
Cold War, 52, 253
collaboration, 283–85
collective mode work, 278–82
collective whispering, 273–87
college, 231
community, 286–87
compassion, 19, 22–24, 57, 80, 81, 92,
    106, 119, 124, 125, 149, 159, 162–64,
    172, 190–91, 192, 267, 271, 281, 290
    awakening, 296–97
    six signs of wise compassion, 185–87
concentration, 169–72, 173, 186
confidence, 72, 80, 150, 182
confusion, 14, 15, 20, 29, 66, 82
connection, 5–7, 12, 18, 48–49, 225,
    235–44, 298
    at the source, 289–98
    authentic, 74–75
    heart whispering, 237–39
    insecure, 37–46
    joined at the heart, 235–44
    joining up methods, 242–44
    patterns of, 5–7
    shared secure base, 245–59
    through the arts, 268–71
    two-way street, 239–40
    virtual, 240–42
control, 52, 54, 136, 138–40, 239
Coordinator, 115
core beliefs, 63–65, 67, 69, 73, 79, 108
    listen of common, 67–68

correctives, mode, 144–47
Costa Rica, 294
couple modes, 224–29
craniosacral therapy, 230
craving, 26, 31, 197, 198, 210
Cromer, David, 7–8
cruelty, 26, 56

Dalai Lama, 10, 38, 57, 77, 129–30,
    134, 159, 162, 191, 197, 232, 253,
    255, 257, 266–67, 271, 290, 291,
    296
danger zone, 137–40
Darwin, Charles, 52
Darwin's orchid, 52
Das, Chitresh, 268, 269
Dass, Ram, 265
Davidson, Richardson, 153, 162
death, 75, 97, 191–92, 233–34, 237
decision, 115
de-linking, 205–7
delusion, 14, 15, 26, 28
"demanding parent," 61
depression, 7, 17, 20, 77, 165–66, 174
deprivation, 67, 79–80
detachment, 40
Dharamsala, 38, 87, 162
diabetes, 140
differences, beneath the, 253–55
discernment, 110–11, 117–20, 124, 135,
    145, 150, 172, 198, 205, 208
disease, 140
dissatisfaction, exploring, 215
dogs, 56–57
dominance, 51
drenpa, 100, 101, 105, 115
drive, 115
drug abuse, 134, 257
dukkha, 26
duty-bound mode, 60–61, 62, 63, 125,
    138
dysfunction, 14, 15
dzinpa, 188

Eastern Europe, 261, 276, 282, 296
East-West fusion, 10–12, 151, 182
Ebadi, Shirin, 257

Einstein, Albert, 14
e-mail, 241
Emerson, Ralph Waldo, 183
emotional empathy, 246, 247
emotions, 8, 134–35, 188
    chocolate mindfulness, 198–99
    clarifying anger, 216–17
    de-linking, 205–7
    deprivation, 67
evolution of, 10–12, 69–83
    finding our strengths, 79–81
    hooks, 203–5
    identifying, 174–77
    physics of, 195–217
    positive, 70–78
    practice guidance, 211–17
    relationship modes, 221–34, 235–44
    shared secure base, 245–59
    signs of progress, 210–11
    tracking the chain, 199–200
    wise heart, 179–93
    wise reflection, 209–10
    See also specific emotions
empathic concern, 247
empathy, 23, 48, 52–54, 56, 70, 119,
    135, 145, 149, 150, 154, 204, 244,
    246–48, 251, 252, 283
    kinds of, 246–47
    pauses, 52–54
emptiness, 208–9
endorphins, 263
energetic flows, 160–62
England, 113, 269, 270, 276
entitlement, 48, 68, 80, 137
Epictetus, 20
epigenetics, 140
equanimity, 18, 19, 77, 152, 181, 216
ethics, 96, 186
everyday habits, 202–3
evolution, 47–57, 182, 297
    of emotion, 10–12, 69–83
empathic pauses, 52–54
    herd hierarchy, 49–52
    predator-prey dynamic, 47–57
    upgrade, 55–57
execution, 115
extreme modes, 62–63, 133, 134

Facebook, 193, 228, 241
failure, fear of, 66, 67
feelings, identifying, 174–77
fight, flight, or freeze response, 33–35,
    52, 62
fish, 297–98
flamenco, 264
flower arrangement, 212, 266
"flow" state, 77
food, 156
    chocolate mindfulness, 198–99
    junk, 209
forgiveness, 65, 274
Frankl, Victor, 270–71
Frederickson, Barbara, 77, 78, 162
French West Indies, 195
fun, 78

Gandhi, Mahatma, 92, 248, 252, 277,
    278, 290
generosity, 186
genetics, 32, 140
gentle path, 248–50
Gilbert, Elizabeth, *Eat, Pray, Love*,
    223
giraffe and the jackal, 250–53
Good Samaritan, 185
Gorbachev, Mikhail, 293
Gottman, John, 221
gratitude, 78
great expanse, dissolving into, 297–98
greed. *See* attachment
Greek Stoics, 182
Green, Paul, 281, 293
guidance, practice, 211–17
gypsies, 261–62, 264

habits, 15–17, 23, 79, 91, 115, 137, 170,
    183, 197, 213–14, 295
    change, 88–92, 108–10, 115, 142–44,
        213–14
    everyday, 202–3
    neuroscience of, 15–17
Hafiz, 211
Hanshan, 183
happiness, 71, 76, 181, 184
harmony, 253

hatred, 26
Hawken, Paul, 297
heart, 171, 179–93
    joined at the, 235–44
    revolution of, 255–58
    whispering, 237–39
    wise, 179–93
helpers, 155–58
herd hierarchy, 49–52
Hesse, Hermann, *Siddhartha*, 151, 193
Hinduism, 182, 277
Hispanics, 282–83
hooks, 203–5
hormones, 225
    stress, 34
horses, 5–6, 21, 49–51, 72, 106–7,
    115–16, 117, 168, 241
    as prey species, 49–51
    whispering, 6, 11, 49–50, 53, 99–100,
        160–61, 191–92, 208, 226, 235–36,
        237–38, 242–44, 248–49, 252–53,
        254, 292
humility, 80
humor, 78
Hungary, 279
hunting and gathering, 49

identifying feelings, 174–77
ignorance. *See* bewilderment
I–It relationship, 55–56, 203, 239, 240,
    252
illness and injury, 72–74, 75, 93–94,
    233, 258
immune system, 297
impermanence, 208, 210
Impressionism, 19–20
impulsivity, 116, 117
India, 10, 13, 87–88, 96, 102, 151, 185,
    261, 262, 265, 275, 276, 296
India Jazz Suites, 268, 269
inflammation, 140
injustice, 26
inner voices, 140–42
insects, 59–60, 195–96
insecure modes, 37–46, 170
    anxious mode, 38, 39, 42–44, 45–46,
        170

avoidant mode, 38, 39–42, 45–46, 170
    See also specific modes
insight, 172–74, 175
    practice, 200–201
Integrity House, Newark, New Jersey,
    257
intention, setting, 258–59
Internet, 241–42
intuition, 119
Iran, 257
I–You relationship, 56, 236, 246

jackal approach, 250–53
Jamaica, 269–70
James, William, 117
Japan, 179, 184, 187, 275
    flower arrangement, 212, 266
Jews, 279
Jinpa, Thupten, 267
Johnson, Samuel, 199
joining up, 6
joy, 4, 78, 80, 96, 181, 269, 296
justice, 57, 71

Kagyu school, 165
Karuna Center for Peacebuilding, 281
kathak, 262
Kerr, David, 257
Khen Rinpoche, Nyoshul, 267
Khyentse Rinpoche, Dilgo, 191
kindheartedness, 5, 274, 276, 289, 290
King, Martin Luther, Jr., 185, 257–58,
    290, 293
King, Martin Luther III, 257–58
kitchen, 158
Kongtrul Rinpoche, Dzigar, 204
Kornfield, Jack, 29
Kyoto, 179, 275

labyrinth, 179–80
Langri, Sophie, 250–51
Latcho Drom (film), 261–62
learning, 123
Libet, Benjamin, 105
life experience, teachings of, 191–93
lightness of being, 187–91, 207
locking, mode, 229–31

London, 113, 276, 296
lotus effect, 3–12, 15, 200
lotus flower, 3–12, 15, 178, 187–88
love, 72, 153–55, 157, 181, 192, 223–24,
    271
    power of, 153–55
    relationship modes, 221–34
    secure mode and, 153–55
lovingkindness, 23, 77, 124, 125, 132,
    162–64, 296
    cultivating, 162–64
lung, 161
lungta, 161–62

Macy, Joanna, 239
Madagascar, 52
Madrid, 122
Magritte, Rene, 4, 207
Mam, Somaly, 256–57
Mandela, Nelson, 290
Marley, Bob, 269–70
    "Redemption," 180
marriage, 138–39, 221–22, 224–26
"marshmallow test," 116
maze, 179–80
McCollum, Ian, 47
McCourt, Frank, 254
McDonough, Bill, *Cradle to Cradle*, 270
meditation, 10–11, 101, 104, 122, 123,
    124–26, 129, 132, 151–52, 162,
    165–78, 183–84, 198, 200, 208, 211,
    274, 278
    arts, 183–84
    calming the mind, 168–69
    identifying feelings, 174–77
    lovingkindness, 162–64
    mindfulness training, 172–74
    one-pointedness, 169–72
    practice guidance, 211–17
    training the mind, 165–78
    vipassana, 200–201, 207, 208
    walking, 106
memory, 94
Mikulincer, Mario, 156, 283
mind, training the, 165–78
Mind and Life Institute, 10, 87, 162,
    267, 296

mindfulness, 16, 61, 72, 90, 100–111, 132, 135, 165–78, 197, 267
  applied, 124–26
  art of whispering, 99–111
  chocolate, 198–99
  collective whispering, 273–87
  of feelings, 174–77
  habit change, 88–92, 115, 142–44
  kinds of, 209
  one-pointedness, 169–72
  overseer, 113–27, 142–44
  physics of emotion, 195–217
  practice guidance, 211–17
  principles of, 208–9
  training, 172–74
  training the mind, 165–78
mind whispering, 3–12, 85–217, 271
  art of, 99–111
  collective, 273–87
  connected at the source, 289–98
  joined at the heart, 235–44
mindful overseer, 113–27
  mode work, 129–47
  physics of emotion, 195–217
  relationships and, 55–56, 221–34, 235–44
  secure base and, 149–64
  shared secure base, 245–59
  shifting the lens, 87–97
  training the mind, 165–78
  wise heart, 179–93
Mingyur Rinpoche, Yongey, 165–67, 177
Mister Rogers' Neighborhood, 156
mode lock, 229–31
mode primes, subtle, 212–13
modes of being, 1–83
  choice and, 95–97
  collective whispering, 273–87
  correctives, 145–47
  evolution and, 47–59
  evolution of emotion, 69–83
  habitual, 15–17, 23
  insecure, 37–46
  lotus effect, 3–12, 15
  mindful overseer, 113–27, 142–44
  perceptions and, 19–20

phase transitions, 17–19
physics of emotion, 195–217
practice guidance, 211–17
predator-prey dynamic, 47–57
priming our secure base, 149–64
recognition of, 21–22, 81–83
relationships and, 55–56, 221–34, 235–44
root causes, 25–35
shared secure base, 245–59
shifting the lens, 87–97
traps, triggers, and core beliefs, 59–69
voices of, 140–42
what is behind, 134–35
work on, 129–47
worlds of, and why they matter, 13–24
See also specific modes
Monet, Claude, 19–20
Mongolia, 229
Motivator, 115
music and dance, 113–14, 158, 261–64, 268–70
Muslims, 277
myth, 96, 97

narcissism, 48, 52, 63–64, 137
Native Americans, 12, 72, 96, 200
nature, 158
Nazism, 276, 279
negativity, 17–19, 30, 31, 33, 60–61, 77, 70, 80, 81, 82, 83, 88, 90, 96, 105, 118, 121, 121, 122, 133, 145, 160, 175, 197, 206, 207, 210, 231, 232, 243, 245, 251, 258, 295. *See also specific modes*
neocortex, 114
Nepal, 10, 281
neurons, 78, 93, 246, 268, 269
neuroplasticity, 93
neuroscience, 11–12
  of habit, 15–17
New Delhi, 87–88, 296
New York City, 56, 177, 185, 263, 264, 275
9/11, 263
Nobel Peace Prize, 290

nonviolence, 255–58, 277
  communication, 250–55
Nyima Rinpoche, Chökyi, 192, 259

Obama, Barack, 284
Occupy movement, 277
"old dogs," 183
One Love Peace Concert, 270
one-pointedness, 169–72
oscillators, 269
outer space, inner space, 211–12
overseer, 113–27
  habit change and, 142–44
  mindful, 113–27, 142–44
Oxford English Dictionary, 169
oxytocin, 154

Padmasambhava, 189
Pague, Steven, 294
Pakistan, 296
panic disorder, 130, 165, 166–67, 171
paramita, 185–87
partners. See relationship modes
passivity, 49, 52
patience, 186
patterns that connect, 5–7
peace, 256, 257, 270, 281–82
Pema Kunsang, Erik, 190
Perceiver, 114
perception, 19–20
  new, 19–20
  wise reflection, 209–10
perfectionism, 60, 61, 68, 79, 80, 83,
    133, 147, 169, 232, 247
perspective, 131–32
phase transitions, 17–19, 77
physics, 208, 209, 297, 298
  of emotion, 195–217
plants, 150
positivity, 18–19, 20, 30, 31, 70–78,
    79, 80, 82, 83, 105, 160, 232, 245,
    258–59, 295. See also specific modes
post-traumatic stress disorder (PTSD),
    94, 159
practice guidance, 211–17
predator-prey mode, 18, 25, 32, 47–57,
    61, 62–63, 80, 81, 83, 136, 137, 140,
143, 147, 168, 239–40, 242–43, 248,
    251, 279–80
prefrontal cortex, 76, 103, 114–15,
    116–17, 121
primal choice, 31–33
priming our secure base, 149–64
progress, signs of, 210–11
psychology, 10, 11, 18–19, 188, 208, 281
  Buddhist, see Buddhism
  developmental, 18
  evolutionary, 18, 47–57, 71
psychosclerosis, 91
"punitive parent," 65
pura vida, 294

qigong, 71, 157–58, 161–62, 197, 294

rabbits, 31–34
racism, 26
Rangoon, 165
reappraisal, 120–21
rebooting, 93–95
recalculation, 107–8
recognition, of modes, 21–22, 81–83
reflection, 123–24
  wise, 209–10
reframing, 144–47, 160
reggae, 269–70
rejection, 41
relationship modes, 45, 55–56, 92,
    221–34, 235–44
  beneath the differences, 253–55
  collective whispering, 273–87
  connected at the source, 289–98
  couple modes, 224–29
  I–It, 55–56, 203, 239, 240, 252
  I–You, 56, 2346, 246
  joined at the heart, 235–44
  mode lock, 229–31
  shared secure base, 245–59
  we are not our modes, 231–34
  See also specific modes
relaxation, 88
religion, 96
resonance, 229–30
revolution of the heart, 255–58
rewards, 89

Rideau, Wilbert, 255–56
Roberts, Monty, 249
Rogers, Fred, 156
root causes, 25–35, 80
Rosenberg, Marshall, 254, 292
Rosenfeld, Arthur, 274
Roshi, Suzuki, 19
routines, 89, 91
Russia, 253, 293, 296
Rwanda, 279, 281

saddhus, 185
Sadowski, Bob, 5–6, 21, 99–100
safe haven, 151–53
samadhi, 170
samatha, 169
samsara, 190, 191
San Francisco, 266
sanje, 4, 15
San Juan Islands, 9
Santa Fe, New Mexico, 267
Sarvodaya Movement, 276–78
sati, 173
satyagraha, 252
schema therapy, 61, 71, 137–39
secure mode, 39, 69–83, 94, 103, 106,
    133–34, 137, 145, 149–61, 182,
    245–59
  collective, 273–87
  cultivating lovingkindness, 162–64
  energetic flows, 160–62
  gentle path, 248–50
  helpers, 155–58
  mind expansion, 159–60
  nonviolent communication, 250–55
  power of love, 153–55
  priming, 149–64
  revolution of the heart, 255–58
  safe haven, 151–53
  shared secure base, 245–59
self-absorption, 22
self-sacrifice, 67, 141, 140–44
serenity, 78
service, spirit of, 265–67
sesshin, 100, 101, 105, 115, 118
seva, 265
Seva Foundation, 78, 265, 266

Shantideva, The Way of the Bodhisattva,
    295
Shaver, Phillip, 38, 45, 156, 162, 283
shifting the lens, 87–97
Siegel, Daniel, 236
six signs of wise compassion, 185–87
slave labor, 187
smallpox, 265
Smith, Jason Samuels, 268, 269
Smithsonian Institution, 200
social exclusion, 68
Socrates, 96
source, connection at the, 289–98
South Africa, 290
Spain, 261, 262
speedy energy, 161
spirit of service, 265–67
squirrels, 56–57
Sri Lanka, 276, 277
Star Wars, 97
Staub, Ervin, 279–80
staying awake, 214
Stoics, 182
Story of the Weeping Camel, The (film),
    229
strengths, 79–81
  finding our, 79–81
stress, 77
  hormones, 34
  post-trauma, 93–94, 159
subjugation, 67, 79
submission, 51
subtle mode primes, 212–13
suffering, 26, 134, 151, 197, 208, 209,
    278
  conditional, 197, 203
suicide, 17, 72
Suu Kyi, Aung San, 200, 255, 269, 290,
    293
survival, 47–57, 72–73
survival of the fittest, 52

t'ai chi, 274
tea bowl, 184, 275
teachings of life experience, 191–93
Teresa, Mother, 290
Thailand, 159, 275

Theravada tradition, 165
Tibet, 10, 123, 124, 161, 162, 177, 182, 183, 188, 189, 191, 192, 203, 211, 275
torture, 56, 110
tracking the chain, 199–200
traffic, 87–88, 94
training the mind, 165–78
trauma, 72–74, 93–94, 159
triggers, 63, 66–68, 79, 89–90, 93, 102, 108, 134, 138, 140, 144, 145, 167, 198, 204, 221, 223, 230, 254
Trungpa Rinpoche, Chögyam, 181–82
trust, 249
Tseringma, Khandro, 167, 207, 238
Tsoknyi Rinpoche, 87, 88, 211, 293
Tutu, Desmond, 290
Twain, Mark, 44
two-way street, 239–40

unconscious, 16, 41, 44, 90, 110, 197
Understanding, 114–15
universal language, 275–76
unlovability, 67, 80
unrelenting standards, 68
unseen forces, 9–10
U Pandita, Sayadaw, 200–201, 255, 267
Urgyen Rinpoche, Tulku, 187, 189–90, 211
U Tejaniya, Sayadaw, 124, 165–67, 174
U Vivekananda, 188, 281

Vajrayana, 165
vedana, 175
Verbalizer, 114–15
vipassana, 200–201, 207, 208
virtual connections, 240–42
visual art, 263

vitality, 186
voices of the mode, 140–42
vulnerability, 68, 80

wakefulness, from bewilderment to, 213–14
waking up, 187–91
walking meditation, 106
Wallenberg, Raoul, 279, 280
wasps, 59–60
water, 209, 284, 297–98
"what-if" cortex, 45
Whyte, David, 61
Williams, John, 113–14
willpower, 109, 168, 174
wisdom, 15, 19, 20, 77, 80, 102–4, 134, 167, 179–93, 207, 209–10, 290
    heart, 179–93
    reflection, 209–10
    six signs of wise compassion, 185–87
wise heart, 179–83
    meditative arts, 183–84
    six signs of wise compassion, 185–87
    teachings of life experience, 191–93
    waking up, 187–91
Wizard of Oz, The, 16
Wolf, Aaron, 284
Wood, Perry, 244
World Health Organization, 265

Yang Yang, Master, 157–58
yee, 197
Young, Jeffrey, 61, 65, 71, 129, 137
YouTube, 156

Zen, 19, 91, 183, 187
Zinn, Howard, 295